Building on Batik

The globalization of a craft community

Edited by
MICHAEL HITCHCOCK
WIENDU NURYANTI

Routledge
Taylor & Francis Group

LONDON AND NEW YORK

First published 2000 by Ashgate Publishing

2 Park Square, Milton Park, Abingdon, Oxon OX14 4RN
711 Third Avenue, New York, NY 10017, USA

Routledge is an imprint of the Taylor & Francis Group, an informa business

First issued in paperback 2016

British Library Cataloguing in Publication Data
Building on batik : the globalization of a craft community.
  - (University of North London voices in development management)
  1. Batik - Indonesia - History 2. Batik - Social aspects - Indonesia 3. Cultural property - Indonesia 4. Tourist trade and art - Indonesia
  I. Hitchcock, Michael II. Nuryanti, Wiendu III. University of North London
  306.4'89'09598

Library of Congress Catalog Card Number: 99-76650

ISBN 13: 978-1-84014-987-6 (hbk)
ISBN 13: 978-1-138-26740-4 (pbk)

Contents

List of Illustrations

Foreword

HELMUT SCHMIDT

Excellencies, ladies and gentlemen, let me thank you in the first place for inviting me, although I am neither an expert of tourism nor am I knowledgeable in the art of batik. I have to admit that it was the topic given to me by the hosts of the conference and that did intrigue me and attract me to visit Indonesia once again.

I have been in your country also on an earlier occasion and indeed my first plans to come to Indonesia I elaborated already in 1939. I was then about to finish my two years of recruited service due to being drafted in the German forces but I was so fed up with the Nazi dictatorship at home, that I decided to join the Shell Oil Company and emigrate to your beautiful country, which was then still a Dutch colony. But before my plans were carried out, Hitler started WWII and I had to serve another six years and never came here.

So my first visit to Indonesia only came about four decades later. And as regards Yogyakarta, this very weekend was the first time I ever saw this city, the Borobudur temple, as well as the Prambanan temple and as well Mt Merapi; not to mention the beautiful fashion show of last night. I must admit I feel overwhelmed by the serene beauty and the invaluable treasure of the two temples.

Now let me address the topic, Cultural Identity in an Era of Globalization, a concept that did not even exist 10 years ago. One of the aspects of globalization is the ever-quickening pace of traffic and transportation. Less than a century ago, one was using sailing boats between India and Java and it took months to sail from Surabaya to Europe. Today there are dozens of jet passenger aircraft loading masses of passengers every day at Jakarta airport. It takes only half a day to fly from Jakarta to Paris, and even less to fly to Beijing.

In other words, millions of people every day are meeting millions of foreigners; the distances are shrinking to a small fraction of what they used to be over centuries, and over millennia.

Another aspect of globalization is the enormous speed of communication. When I was a boy, one used to write postcards or letters and a letter from

Hamburg to Batavia took more than one month. Today, you don't only have telephones but faxes by satellites. If something happens on the stock exchanges in Hong Kong or Bangkok, within seconds it creates an effect in Jakarta, Tokyo or New York.

This shrinking is the effect of an enormous development of technology whether in transportation or telecommunications. Plus, the new financing techniques; for example, today you can buy Japanese yen-denominated shares electronically and sell the paper again the same day in New York.

The new phenomenon in the last quarter of this century is the high speed of technological progress plus the high speed with which new technologies are freely disbursed and exported, and accepted as well all over the globe.

This is true in all fields, in commerce, in finance, or manufacturing, or military weapons, or in medicine, or in science as well as in television. This is a very high speed of dispersion of new technology that never happened in former centuries. For instance, it took the steam engine more than a century to spread from England to China. It took the passenger aircraft just a quarter of a century after it was developed in the West to be produced by your colleague Prof. Habibie in Indonesia. The compact disc or the cellular telephone needed less than a year to be spread all over the world.

In addition to technological globalization, the number of participants in the global economy has doubled over the past 20 years or so. For instance, all these 16 successor states of the Soviet Union that had been sealed off from the world's economy, all of them have become open states nowadays participating in the world's economy.

Moreover, in addition to this doubling of participating people and states, almost all of them have opened up their borders to a higher freedom of traffic and trade then ever before in history. We are today experiencing the highest degree of liberalisation that the world has ever seen.

Of course, economic globalization has offered great chances for developing countries, for low-wage countries, and never before have the nations in Southeast Asia enjoyed such a speedy economic progress, and such a speedy increase in their standard of living in real terms – never, ever before.

On the other hand, in the old manufacturing industries in Eastern Europe, this same process has led to a loss of jobs. We are losing our jobs to Indonesia, for instance, and to other countries in Southeast Asia as well. The economic growth rates in Southeast Asia have been speeding up and the growth rates in Europe have gone down. This is the inevitable consequence [of globalization], as if globalization is only helping the Southeast Asians.

But globalization also entails severe economic risks as well – financial risk, monetary risks in the quick growing economy of Asia. The present currency crisis in Thailand or Indonesia is the latest illustration of what I am saying. It is once again obvious that great chances do go hand-in-hand with great risks.

But then economic risks are not the only risks that are inevitably created by economic, technological, and communication-driven globalization. There exists as well psychological risks, there do exist educational dangers, and there does exist a threat to the cultural identity of whole societies and whole nations.

Psychological or even ideological adaptation of many businessmen, for instance, whether in Europe or Asia has presently led to an almost globalized trend towards speculation, and towards predatory capitalism. There is a growing temptation of financiers for their overriding purpose in their lives to become as rich as possible and as soon as possible.

This trend started in America, secondly it infected Japan, then the so-called four tigers, and at the same time it affected Europe and great parts of Southeast Asia, including Hong Kong and Shanghai. Yesterday, the Indonesian government had to close a dozen or so private banks. This is only mirroring the results from greedy and all-too-risky financial behaviour that is today spreading all over the globe.

We do not have a universally accepted code of morality, as long as we do not have a minimal code of ethical behaviour, it seems to me that it remains a task for the nations and their governments within their states to uphold and maintain the ethical behaviour of businessmen.

It is obvious that this task is made easier and more successful the more a society and the more their political leaders stick to the ethics they have inherited from their forebears, from their religions, and from solid cultural traditions all together.

Another aspect of globalization is the danger of mis-education of the young, and mis-education of whole societies by way of globalized television and media. The more wealthy a nation gets, the greater the available number of international television channels and the greater the number of TV sets in private homes. In my country, for instance, children between 10 and 14 years of age, on average spend 3–4 hours watching television every day. Television has become a major factor in their education whether we like it or not, and I dislike it heartily.

Because what these German youngsters are exposed to on television every day contains a great deal of crime, murder, abductions, shoot-outs, and violence in all its forms, including rape, plus a lot of cheap sex. The youngsters see it with their own eyes, they hear it with their own ears, and they are made to believe that this in reality is what life is all about. This I call 'mis-education'.

It seems to me that a great educational effort is needed in order to balance out the negative effects of globalized cheap television.

Globalization is at the same time encroaching on our national languages. It seems inevitable that English is quickly becoming the universal language of business but also our everyday private language does contain a growing degree of English words and American slang.

The Italian language has been developed over a long period of history, perhaps 2,500 years. The French language needed some 12 centuries; the German language of today took almost the same time to develop. By contrast, Bahasa Indonesia, the common language of a nation of almost 200 million people, is a rather young language. But however old or however young our language traditions may be, most of them are nowadays being influenced by English injections (vocabulary and grammar alike), by amputations and by overlays at the same time.

Personally, I do hold the view that language is one of the roots of one's cultural identity. We therefore have to be careful to shield our languages against corruption from the outside. In Europe, we are quite aware of such dangers. All the nations who are participating in the enormous undertaking of forming a European Union, an undertaking that started 47 years ago, all of these nations stick to their national languages and stick to their national heritages.

On the one hand, we do understand the necessity to join our capabilities, to form a common market, to jointly stand up against the superpowers of the twenty-first century but none of us will sacrifice his or her national cultures. We will not sacrifice our cultural identities.

I mentioned *Bahasa Indonesia* [Indonesian] and to me it is one of the most impressive parts of the cultural identity of our host country. But then even more impressive to me is *Pancasila* [Indonesian national code], particularly the first of the five principles – religious tolerance. Tolerance not out of neglect or disinterest, but tolerance out of respect for other citizens and their religious beliefs and traditions.

The respect for different religions in Indonesia, all of which have traditions and histories behind them of millennia of years, this respect I think is one of your greatest achievements. It took us Europeans several centuries to reach that degree of respect for other citizens' religions but it took Indonesia much less than half a century. To maintain that respect in my view is one of your greatest responsibilities in the future as well.

To spread religious respect, to spread tolerance beyond your own border,

is a universal necessity if we desire to avoid a so-called 'clash of civilisations' which Harvard Professor Huntingdon is threatening us with. Respect between Islam, Hinduism, Buddhism, Christianity, Judaism, and Confucianism.

Respect is an overriding necessity if we want to preserve peace between a growing number of human beings; in other words, if we really want to preserve peace despite a shrinking space for all living human beings. On the other side of the coin, without any doubt, is the necessity to stick to one's own roots, to one's own religion, one's own behaviour, and one's own cultural heritage.

The French 200 years ago developed the division of power within the state; the Italians and the Germans developed classical music, they have stuck to it and mankind have benefited from this. But then, the people of Indonesia have developed the music of gamelan, the art of batik, the masks, and the shadow play. If you stick to it, and I hope you will, it will help to maintain your cultural characteristics, your identity, despite globalization, and the world outside will enjoy it.

The first time I heard gamelan music was in Amsterdam, and that was 40 years ago. I became interested in hearing more of it even then. As we have heard from the Vice President and the other speakers, batik from Indonesia has become famous all over the world, and the art and technique of batik has spread to many other nations, also to Germany. For instance, at home, my wife and I have three large pieces of batik in our dining room and the studio.

Don't clap your hands too soon! Although they have been done in the traditional way, they have been done by a German artist!

It seems only natural that the temples in Java, the primeval landscape in Kalimantan and many islands, the beauty of the Balinese, it is natural that gamelan and batik do attract visitors and tourists from all over the world. Tourism is of course benefiting the host country, and mass tourism will grow further. It is one of the positive effects of globalization. I do expect it to contribute even more to the economic wellbeing of the Indonesian people.

Any of the 200 nation states on this globe is to a growing degree becoming part of economic, technological, and communicative globalization. Indonesia is the fourth largest of these 200 states but any of these 200 nations would be wise if they stick to their heritage.

If they develop their heritage further, it should only be gradually, step by step, gradually and with great care. If one does this, then all the other countries outside as well will benefit from your cultural achievements. If all of us learn to mutually respect our different cultures and civilisations, then globalization will not endanger our identities.

Let me sum it up by saying the greatest challenge of globalization is twofold. Number one, to heed and shield our own cultural identities, and number two, to respect the cultural identity of any other culture.

This is why we must look to our own cultures and religions. There is one basic thought in all the great religions, which the theologians call the 'golden rule'. That is, 'you do unto others that you wish them to do unto yourself'. If we make this the cornerstone of global culture, I would be satisfied. But it has to be brought to the attention and conscience of Christians, and Muslims, and Hindus and so on. If you look at the Middle East today with fights between Arab Muslims and Jewish Israelis, it seems we are still somewhat away from the global culture of the golden rule.

I think that at least partially the answer lies in the field of education. If you study the curriculum for business administration degree whether it is in Zangal in Switzerland or Harvard, Mass., I would like to have included in the curriculum some ethical teaching. You will not find it there; you will not find any philosophical teaching in the curricula today.

But education influences the young ones, young in their minds and their souls, whom you can influence rather than the older ones of 60 or more.

But if you talk of global culture, the cornerstone of my belief is the imperative to everybody to not behave vis-à-vis others except as you desire them to behave vis a vis yourself.

As regards the elder ones, the grown up businessmen, I feel within society that people should not show any respect to somebody because he is rich or because he has made a fortune in commerce or banking. I think one should pay respect to those who use their riches for good purposes. The old Romans had a proverb: 'salus publica, supremus rex' – 'the public's sanity and well-being is the highest law for anybody'. If this kind of thinking would take over the minds of you and your colleagues in business, all of us would be better off.

You refer to the principle of unity out of diversity. I think you have to carefully analyse where this principle applies and where it doesn't. Of course the Indonesian nation of 200 million is composed of people of many different historical backgrounds and heritage. You have I don't know how many languages between Sulawesi and Sumatra. You have tried at least for daily use in the schools, and universities and administration to replace them with Bahasa Indonesia. Unity out of diversity – good, very good.

The same thing has happened in most of the European nations. We Germans have been originally partially North Germanic tribes, partially Slavic tribes, and partially Celtic tribes. The same is true of England – Celtic, then Normans by way of France, the Anglo Saxons and so forth. In the course of centuries, all these influences whether in Britain, or Germany, or Italy, or in Spain or in France have melted together. But I would try to maintain the case of unity that has developed over 1,000 years – in the case of England, since 1066; in the case of France, since Charlemagne; for Germany, a little later.

We should try to maintain these unified cultures and not try to mix it up with quite different cultures as they have developed for example in Kenya, Somalia, and Zimbabwe. And not to mix up the cultures as they have developed in Zimbabwe with the cultures as they have developed in Brazil; and not to mix up the Brazilian culture with that of China, or those with Bangladesh, or with that of Indonesia. If we allow a hodgepodge, a mix of all to take over, then most of our people will lose their orientation.

Religion and culture as a whole, even if we are not very deep believers, is a root of the human being. Culture does not exist in animals – they have instincts. Culture has been learned in the course of human life; you get acquainted with the culture of your parents, of your neighbourhood, of your school, of your city, of your nation. This is what distinguishes human beings from orangutang or chimpanzees.

If we mix all our culture into one pot, I think it will smell ugly. I think we should stick to our culture and find out if our neighbour has something to offer that we can take from. If he offers us batik shirts like this, I would rather have his shirt than my suit that is much warmer. If he offers something I like, I will take it but I will not get under his thumb and force him to take anything. Let me address the American ambassador, there are some Americans who think what is good for America also is good for the rest of the world. This is a principle I would not buy.

Transcript of Chan. Helmut Schmidt's Keynote Address
Dunia Batik Conference and Exhibition
Yogyakarta, Indonesia
November 1997

Acknowledgments

This book grew out of the *Dunia Batik* (World Batik) Conference which was hosted by Gadjah Mada University in Indonesia 2–5 November 1997. The theme of the conference was 'The Role of Textiles in Tourism and Constructing Cultural Identity'.

The symposium took place in Yogyakarta and included the following associated activities: 'Spirit of the *Selendang*' competition; Batik Youth Forum; Charity Night for Community Projects in Batik Villages in Central Java; and the *Dunia Batik* Exhibition. Parallel presentations and workshops were held in the following cultural and historic sites in Yogyakarta and Central Java: the Afandi Museum, Yogyakarta; Gadjah Mada University; *Keraton Kasunanan* (Sultan's Palace), Solo; and Ndalem Tjokrosumartan Emporium, Yogyakarta.

The editors gratefully acknowledge the following for their unstinting support for the conference: Ministry of Tourism, Post and Telecommunications, Republic of Indonesia (RI); Ministry of Education and Culture, RI; Ministry of Trade and Industry, RI; International Centre for Culture and Tourism (ICCT), Yogyakarta; National Museum of Indonesia, Jakarta; Indonesian Textile Association, RI; Indonesian Batik Foundation, RI; and the Institute for Research and Development, Handicrafts and Batik Industry. The editors are especially grateful to H.E. Joop Ave, former Minister of Tourism, Post and Telecommunications, RI, whose generous support was vital in ensuring that the conference took place and enjoyed such widespread international support. Heartfelt thanks are due to Helmut Schmidt, former Chancellor of the Federal Republic of Germany, for his opening address, participation and inspiration.

We appreciate the support of Professor Margaret Grieco, the Voices in Development series editor, Pat FitzGerald, who prepared the camera-ready copy, and the Business School of the University of North London. We would also like to thank all the participants and exhibitors at the *Dunia Batik* symposium for making the conference such a success. Finally, we shall always be indebted to the staff and students of Gajah Mada University for their warmth and hospitality at a time of momentous change in the history of Indonesia.

Michael Hitchcock and Wiendu Nuryanti

Introduction

MICHAEL HITCHCOCK AND WIENDU NURYANTI

Batik is a method of dyeing cloth in which the waxed parts are left uncoloured. The beauty of batik bears witness to the patience, creativity and industry of the craft communities involved in the production of decorative textiles. Batik designs and materials vary from place to place, and it has been estimated that there are well over 3,000 batik designs in existence. It is an intricate and labour-intensive textile process, a mainstay of both rural and urban workshops in the developing world. Batik has folk as well as royal origins, and has inspired artists and designers worldwide and is in much demand as a fashion accessory. It is a versatile craft that can easily be adapted to the needs of educationalists, hobby craft enthusiasts and tourists alike. It is these qualities that have led to the emergence of batik as one of the first global handicrafts.

The word batik is of Malay-Indonesian origin and is thought to derive from the term *tik* meaning to 'drop' or 'drip'. This suffix may also be detected in other Indonesian words, such as *teritik* (a technique involving sewing and gathering the cloth to resist the dye), *klitik* (name of a batik design) and *nitik* (batik patterns which imitate woven designs) (Fraser-Lu, 1986, p. 1). The Javanese are widely regarded as the most renowned makers of batik, though they are not the only people to use this technique. Batik, for example, was used in Egypt in the fifth and sixth centuries AD, and there are well established batik traditions in China, Japan, Central Asia and West Africa. The term batik is not mentioned in Old Javanese, and this has led some scholars to conclude that the technique was not known early in Javanese history. The absence of the term 'batik' in Old Javanese may, however, be due to the fact that it is possibly of Malay origin. The word *tulis* was used to refer to certain kinds of cloth in Javanese texts from the twelfth century (Wisseman Christie, 1993, p. 16), and the term was later applied to the process of drawing in hot wax with a pen known as a *canting*.

Early texts from Java reveal the importance of cloth in the symbolic and material culture of the island. References to textiles in the literature based on Indian models remain vague, but in legal documents, notably the records of tax transfers known as *sima*, the references are more numerous and specific

(Wisseman Christie, 1993, p. 11). The latter documents, preserved on stone or copper plate, record the transfer of tax or labour rights by rulers or high ranking officials. Different kinds of fabrics were presented on these occasions, and textiles feature prominently in the gift lists associated with these charters. During the tenth and eleventh centuries the charters began to supply more information on taxable activities, but by the middle of the eleventh century the emphasis had switched to increasingly complicated regulations concerning the regalia of rank and restrictions of the use of certain kinds of cloth. After the tenth century more emphasis was placed on gold coins and the importance of cloth as ceremonial gifts declined, though *sima* documents increasingly began to list privileges, including the right to wear certain kinds of cloth (ibid., p. 12).

Javanese textiles were mainly produced to satisfy local demand, though they were produced in sufficient quantities to attract foreign interest. Envoys from Java took textiles as gifts to China, and there is evidence that cloth was used in trade. Chau Ju-Kua, for instance, a Chinese harbour official dealing with overseas trade, referred to Java in 1225 as a source of various coloured brocades, gauzes and damasked cotton. He also mentioned sericulture (Chau, 1966). Ma Huan, a fifteenth century Muslim Chinese chronicler, also noted the use of silk, notably the patterned silk sarongs worn by the ruler of Java (1433).

Ma Huan's account indicative of the growing strength of Muslim traders, the religion having arrived in the archipelago perhaps as early as the eighth century. Islam was introduced via the southeast corner of India to Sumatra, whence the religion spread eastward to the coastal areas of Indonesia. Muslim merchants were interested in local products such as spices and sandalwood, though some doubtless became involved in trading cloth.

Seeking direct access to the Spice Trade, the Portuguese rounded the Cape of Good Hope and entered the Indian Ocean. They were followed by other European mercantile powers, notably the Dutch and the English. The Europeans also shipped cloth, and became involved in the *patola* trade between Gujarat and Southeast Asia. Indigenous patterned textiles were also traded and there are references to a shipment of *kembangan* – possibly Javanese *teritik* cloth – in 1580, and to 'batick' in 1641 on a bill of lading for cloth shipped from Batavia (Jakarta) (Gittinger, 1990, p. 16). Locally produced substitutes for prestigious Indian cloth were also developed in the seventeenth and eighteenth centuries (Maxwell, 1990, p. 236).

Central Java

There are three major batik areas in Java – central Java, southwest Java and the north coast. The old sultanate capitals of Yogyakarta and Surakarta, both of which are famed for their fabrics, are located in the first of these regions. Yogyakarta dyers are known for their strong designs set in *soga* (chocolate brown) and dark blue against a white background, whereas Surakarta's are distinguished by their use of *soga* on a pale yellow background. The batik designs of Central Java are drawn from a broadly based Javanese tradition, though the royal courts in the eighteenth century issued ordinances forbidding commoners from wearing specific patterns. Both cities are centres of traditional as well as commercial batik production and have numerous family run businesses. Surakarta is, however, the home of Batik Keris, a company employing 8,000 workers with retail outlets throughout Java (Fraser-Lu, 1986, p. 60).

Family run concerns also play and important role in southwest Java where *tulis* is prepared alongside batik *cap* (stamp) and screen printing. The batiks resemble those of central Java, particularly with regard to the blues and browns, though there are marked differences. Various reds, greenish-blues and black are used in Tasikmalaya, whereas in Banyumas russet and golden brown on a dark blue background is popular. Small factories in Garut and Ciamus produce batiks that resemble those of Tasikmalaya (Fraser-Lu, 1986, pp. 62–3).

North Coast Java

The batik makers of Java's north coast, a region stretching from the Straits of Sunda to Banyuwangi, have long been exposed to foreign influences. The population is more heterogenous than inland Java, and among the batik entrepreneurs are many people of mixed and Chinese origin. Pekalongan, for example, became well known for its European and Chinese batik compounds and their imaginative design work (Vuldy, 1991, p. 171). Over several generations, at least until the nineteenth century, the Chinese tended to assimilate with the indigenous population. These communities – known as Peranakan in Indonesia, and Baba or Straits Chinese in Malaysia and Singapore – spoke local languages but never entirely lost their sense of Chinese identity.

They settled in cities such as Pekalongan and, though initially regarded with suspicion by the Central Javanese, eventually became well established in trade. By the late nineteenth century the Perankan Chinese families had become involved in the management and distribution of batik throughout much of Java, and each north coast centre evolved its own distinctive style (Maxwell, 1990, p. 264). The Chinese of Ceribon, for example, developed a style of batik known as *kanduran*, which was named after the district settled by the Chinese (ibid., p. 265). Chinese inspired designs were incorporated into north Javanese batik, such as the *lokcan* (combined bird and foliage), the *chi-lin* (Chinese unicorn) and the red and blue cloud motifs of Ceribon. *Peranakan* women, though they adopted the Javanese style blouse and sarong, could readily be distinguished by their style of batik.

Pekalongan became a popular place to live for the less wealthy Europeans and Indo-Europeans. It was cheaper than Batavia, as well as cooler, and the woman could work in the batik trade (Veldhuisen, 1991, p. 165). From about 1860 onwards, north coast batik entrepreneurs adopted the European convention of signing their works, breaking with the indigenous custom of anonymity.

European Batik

The Europeans admired the skill of Javanese batik makers, and began to document the craft and acquire collections. Some of the oldest surviving batiks were made to order for European textile manufacturers eager to investigate what kinds of cloth were popular in Asia (Maxwell, 1990, p. 369). The British Museum, for example, holds batiks that were collected by Sir Thomas Stamford Raffles when he was Lieutenant Governor General of Java (1811–15). The British unsuccessfully attempted to sell British made fabrics in Java, and this may have inspired them to investigate Javanese methods in an attempt to produce cheaper copies. The British were, however, unable to replicate Javanese dyes and had to use large numbers of rollers to imitate handmade designs, and the costs eventually proved prohibitive. The British did, however, manage to make inroads and in 1820 Crawfurd was able to report that British goods were competing successfully with Indian and Javanese goods: 'The bandana handkerchiefs manufactured at Glasgow, have long superseded the genuine ones [from Java]' (1820, III, pp. 504–5).

The Dutch, who also attempted to manufacture batik, tried a different approach and brought skilled batik makers to the Netherlands as instructors. A batik factory was established in Leiden in 1835, and the idea caught on elsewhere in Europe. The Swiss, for example, began to export imitation batik, but following the adoption of wax block, cap, in Java in the 1870s, Swiss production declined. The Germans also experimented with batik, and by the early 1900s were mass producing batiks with glass pens and electrically heated resist (Dyrenforth, 1988, pp. 14–15).

European artists also appreciated the decorative qualities of batik, and Javanese influences are detectable in art nouveau, particularly in the Netherlands. Batik inlays in furniture were the hallmark of a group of Haarlem based designers. One of the group, Madam Pagnon, opened a small factory in Paris in 1916 (ibid., p. 15). In the 1930s batik was adopted as fashion accessories by tourists and the fashion spread. In *Big Money* (1936), for example, J. Dos Passos writes about 'ladies in flowing batik'. Britain re-entered the batik trade in the aftermath of World War II, producing copies of Javanese batiks for export within the Commonwealth. A collection of these fabrics is housed in the Whitworth Art Gallery in Manchester.

The Dutch colonial authorities also took an interest in the batik trade and made various attempts to stimulate the industry in Java. Extensive research was also conducted on indigenous handicrafts leading to a series of publications that are still the starting point for textile scholars. One of the most comprehensive volumes on batik ever produced, *De Batikkunst* (1916), was jointly authored by a Dutchman, J.E. Jasper, and an Indonesian, Mas Pirngadie.

The aesthetic of the Far East retained its hold on the European imagination until the outbreak of the First World War. The 1909 debut of the Ballet Russes, with costumes by Leon Bakst, successfully scandalised audiences with the so-called *joupes culottes* (harem skirts) in rich, textured fabrics. Harem imagery remained popular throughout the 1920s, and fashion writers continued to praise 'Oriental' designs for their 'exotic', 'vivid', 'dangerous' and 'rainbow' properties (Steele, 1985, pp. 227–34).

Batik and Fashion

From the outset fashion provided an important stimulus for the batik industry. Indian chintz was worn by Europeans in Batavia, but with the stagnation of this trade at the beginning of the nineteenth century, north coast batik, with its chintz-like patterns, came into its own (Veldhuisen, 1993, p. 27). The batik sarong worn with a long *kebaya*, which stayed in fashion until around 1870, became the formal wear of Indo-European women. Male formal wear remained primarily European, though at home or with close friends men wore either sarongs or batiked pyjama trousers. The batik sarong temporarily fell out of favour during the Raffles interregnum: the British regarded the women's attire as inappropriate. Impelled by strong notions of propriety, British women tried to persuade Batavian women to adopt European styles of dress, but this was resisted by Indo-European women. The struggle over what was considered appropriate female attire was not concluded until 1840 when the wife of the Dutch Governor General, de Eerens, introduced European dress for official receptions and visits to the theatre (ibid.). Thereafter, batik disappeared from formal fashion wear.

Batiks, however, continued to be in demand among the European population, which had started to grow in size. The opening of the Suez Canal in 1869 brought the Dutch East Indies closer to Europe in both travelling time and the imagination, stimulating interest in Indonesian products (Veldhuisen, 1993, p. 76). Although the majority of Europeans were men, the numbers of women were significant and had an appreciable impact on batik design. The sarong worn by these women became more European in character, but did not lose their essential Javanese character and retained the distinctive *badan* and *kepala* format.

Bandung, which was just a small outpost at the start of the nineteenth century, gradually developed into an elegant and fashionable centre. It was not just European women who adopted fashionable batik dress, but also local Sundanese women. The multi-hued sarongs from Java's north coast were combined with floral *kebayas* made from Voile de Paris. Batiks drawn from throughout Java were displayed in Bandung after 1900, attracting traders from elsewhere in the Dutch East Indies, as well as from British Malaya, Thailand and Burma.

Tastes began to change, however, and towards the end of the nineteenth century Dutch women began to stop wearing the sarong and *kebaya*, even as informal attire. Increasingly the kimono and *bebe* (a light, loose dress) began to replace the sarong at home, the former becoming popular with Eurasian women as outdoors wear. With the rise of nationalism in the early twentieth century, indigenous demand for batik also began to change. Islam became closely associated with the desire for autonomy and increasing emphasis was placed on Islamic codes of dress to distinguish the *pribumi* (indigenous) from the Europeans and the Chinese. In the 1920s the batik sarong, providing that it refrained from figurative designs, became the preferred clothing of the mosque.

Batik and Tourism

As the European population in the Dutch East Indies grew at the height of the imperial era, so did the demand for local holidays. In 1908 the colonial regime opened its first official tourist bureau in Batavia with the aim of promoting the Dutch East Indies as a holiday destination. This was also the year that the Dutch completed the acquisition of Bali, a process that had begun in 1846, but was not achieved until the bloody campaigns of 1906–8. Rather than accept colonial rule, the last remaining independent Balinese dynasties chose to fight to the death; Holland's international image was severely tarnished by this unnecessary slaughter. In order to atone for their behaviour and to restore faith in their colonial aspirations, the Dutch began to present a more positive image based on the preservation of Balinese culture and the development of tourism. This was also a pragmatic move since newly annexed Bali appeared to offer few commodities suitable for colonial enterprises. The tourism bureau extended its coverage to Bali in 1914, but it was not until the introduction of a regular steamship service from Java in 1924, that tourism took off (Picard, 1993, p. 75)

It was not only the domestic colonial market that fuelled the growth of tourism, but also the growing significance of Bali as an international destination. Opportunities for travel, especially for the upper classes, increased rapidly in the early decades of the twentieth century. The influx of well healed foreigners provided a boost for the makers of handicrafts, and led to the foundation of the Bali Museum in Denpasar under the direction of the German-Russian expatriate painter, Walter Spies. Batik regained some of is earlier popularity as a fashion accessory, particularly among the bohemian hoteliers and their guests in Bali.

Batik and Nationalism

The Dutch empire collapsed in 1942 with the Japanese invasion. Batik designers had to adapt to changing conditions, and Japanese taste is detectable in some of the design of the period. For example, a shortage of cloth, coupled with the need to provide work for artisans, led to the development of *Hokukai* batik with bright floral patterns set on complex *isen* backdrops (Fraser-Lu, 1986, p. 76). The invaders needed local support for their military ambitions and some Nationalists were provided with training, which was later to prove useful in the independence struggle. 'Indonesian resistance' was more of a frame of mind than a movement at the time, and while some cooperated with the Japanese (e.g. Sukarno and Hatta), others went underground but while keeping in touch. With the Japanese facing defeat in 1945, the Nationalists, led by Sukarno, proclaimed independence, though it was not until 1950 that the Dutch were finally ousted.

The Indonesian nation is a deliberately created reality, derived from the Herderian brand of nationalism in which the country's foundations are its people, its common language, the single nation and common traditions (Hubinger, 1992, p. 4). Sukarno encouraged the newly independent Indonesians to look to their own traditions, and engaged the services of K.R.T. Hardjonogoro, one of the leading experts on Javanese culture, to develop a more nationalistic style of batik. As a scholar and artist, Hardjonogoro, attempted to revitalise Central Javanese batiks by combining it with brighter colours from the North Coast (Fraser-Lu, 1986, p. 76). Sukarno also engaged designers to remodel the sarong and *kebaya* on nationalist lines, and had a penchant for celebrating special events with new batik patterns.

President Suharto has also continued in a similar vein, encouraging men to wear batik shirts as formal wear. His wife, the late Siti Hartina (Tien Suharto), who was raised in the Mangkunagoro Palace, was known for her love of the soga batiks of Central Java, and her support of the workshop of Princess Ayu Harjowiratmo.

Tourism, Fashion and Batik

Batik production revived during the post war recovery, and by the summer of 1958 women in countries such as Britain were wearing dresses made of this material. The strong growth of tourism combined with the demand for leisure wear led to the creation of the Hawaiian shirt with its Indonesian references. This essentially American product was to become one of the enduring leitmotifs of tropical tourism. The batik sarong also re-emerged as a popular tourism product in the 1960s and 1970s, becoming an essential part of beachwear, particularly in Australia and California.

By the early 1990s batik had become more than a simple travel accessory, and began to penetrate new arenas of fashion for the first time. The Turkish designer, Rifat Ozbek, was one of the main people behind this transformation, employing new materials such as Lycra to explore the versatility of the sarong. Using motifs drawn from the court tradition, Rifat Ozbek, revisited the enduring design strengths of Central Javanese batik on the international catwalk.

Contemporary interest in batik is not solely the preserve of haute couture, there being a strong popular tradition, especially with regard to souvenirs. In addition to fashion garments, batik has been used to make bags, paintings, belts, kimonos, hats, curtains, upholstery fabrics, tablecloths, bed-linen, wallhangings and lampshades. These goods are not simply purchased on account of their design qualities, albeit a highly significant component, but for a range of other reasons.

Textiles purchased on holiday are often among the most valued possessions. These handicrafts are meaningful and are often more than simple mementos of time and place. Items acquired during a holiday are associated with the travel experience, but are also linked to a generalized image of a culture, or even a specific town or village. Many tourists read around the subject to find out more about their purchases, and some become specialist collectors and experts. Specialist tours involving handicrafts form a small but increasingly important niche market. Other tourists are oriented towards clothing crafts, wearing textiles and ready to wear garments as expressions of taste and identity.

A popular misconception is that cultural products such as batik are historically stable. It is often assumed that there are crafts which may be regarded as traditional and that these are denigrated following the introduction of tourism, as if tourism was the only active agent. Changes in material culture can occur for a variety of reasons. Batik makers have a long tradition of making goods that appeal to a variety of ethnic groups, as is particularly the case with the north coast of Java. Historically, batik was made with not only the highly conscious domestic market in mind, but also the important export trade. Changes in Javanese batik design can be detected long before the advent of tourism, and its is not easy to divide 'touristic' from 'traditional' fabrics. As can be seen from the growing popularity of books on Southeast Asian handicrafts, tourists themselves can be discerning buyers. Many of the people involved in producing goods for tourists, as is especially the case around the court of Yogyakarta, are themselves traditional artisans who have simply adapted their products to suit the tourist market.

References

Chau Ju-Kua (1966), *His Work on the Chinese and Arab Trade in the Twelfth and Thirteenth Centuries Entitled Chu-fan-chi*, trans. and annot. F. Hirth and W.W. Rockhill, New York, Paragon Book Reprint Corp.

Crawfurd, J. (1820), *History of the Indian Archipelago Containing an Account of Manners, Arts, Languages, Religious Institutions, and Commerce of its Inhabitants*, 3 vols, Edinburgh; reprinted London, Frank Cass & Co, 1967; Kuala Lumpur, Oxford University Press, 1967.

Dyrenforth, N. (1988), *The Technique of Batik*, London, B.T. Batford.

Fraser-Lu, S. (1986), *Indonesian Batik: Processes, Patterns and Places*, Singapore, Oxford University Press.

Gittinger, M. (1990), *Splendid Symbols: Textiles and Tradition in Indonesia*, Singapore, Oxford University Press.

Hubinger, V. (1992), 'The creation of Indonesian national identity', *Prague Occasional Papers in Ethnology*, 1, pp. 1–35.

Ma Huan (1433), *Ying-yai Sheng-lan* [*The Overall Survey of the Ocean's Shores*], trans. and ed. Feng Ch'eng-Chun with introduction by J.V.G. Mills, 1970, Cambridge, Cambridge University Press.

Maxwell, R. (1990), *Textiles of Southeast Asia: Tradition, Trade and Transformation*, Melbourne, Oxford University Press.

Picard, M. (1993), 'Cultural tourism in Bali: national integration and regional differentiation' in M. Hitchcock, V.T. King and M.J.G. Parnwell (eds), *Tourism in South-East Asia*, London, Routledge.

Steele, V. (1985), *Fashion and Eroticism: Ideals of Feminine Beauty From the Victorian Era to the Jazz Age*, Oxford, Oxford University Press.

Veldhuisen, H.C. (1991), 'European influence on Javanese commercial batik' in G. Volger and K. von Welck (eds), *Indonesian Textiles*, Cologne, Ethnologica.

Veldhuisen, H.C. (1993), *Batik Belanda 1840–1940: Dutch Influence in Batik From Java History and Stories*, Jakarta, Gaya Favorit Press.

Vuldy, C. (1991), 'Pekalongan as a centre of batik trade' in G. Volger and K. von Welck (eds), *Indonesian Textiles*, Cologne, Ethnologica.

Wisseman Christie, J. (1993), 'Ikat to batik? Epigraphic data on textiles in Java from the ninth and fifteenth centuries' in M-L. Nabholz-Kartaschoff, R. Barnes and D. J. Stuart-Fox (eds), *Weaving Patterns of Life*, Basel, Museum of Ethnograph.

Dunia Batik Conference Declaration, Yogyakarta, Indonesia, 5 November 1997

The origins of batik are diverse in origin, but over time this ancient art has become closely associated with the identity of Indonesia. Since Independence the government of Indonesia has fostered the development of batik as a symbol of the nation's creativity and self-reliance. Today this Conference is proof of the continuing efforts of the government of Indonesia to ensure batik's place as an expression of a living culture, an export commodity, and a tourism attraction.

This goal of this declaration is to affirm Yogyakarta as a world centre for the production and study of batik, and as a starting point for the discovery of the textile arts of Indonesia and the developing world. This is the first major international conference on textiles and tourism, and thus establishes Yogyakarta as the leading centre in this field.

Therefore this Conference affirms:

1. Batik is an art form as well as a social, cultural and economic treasure. It is also a sacred symbol and generally associated with health and well being.
2. An analogue may be drawn regarding batik with traditional medicine being used for commercial purposes by international companies. Therefore, research is needed with regard to intellectual property rights and benefit sharing on how best to protect the rights of traditional batik designers without stifling creativity or preventing the spread of cultural awareness.
3. Creativity is a fragile quality that makes a major contribution to human welfare. The welfare of batik producers from workers to managers and designers must also be a social priority in any development and promotion of batik. Efforts should be made to bring them closer to economic stream.
4. Batik is a cultural art form handed down from generation to generation. In order to ensure its continued high regard with each new generation, it is recommended that Indonesia's best living batik artists, designers and teachers be acknowledged by the country as 'national treasures' following

the Japanese idea. These artisans should be established subject to the strictest criteria to enhance Indonesia's position as a centre for cultural conservation and innovation.

5. Indonesia is a country with one of the world's richest endowments in terms of biodiversity. Historically these islands were a major source of natural dyes. Specialist collectors and tourists seek out fabrics dyed with traditional methods; however, small-scale production is often not viable. Therefore, research into how traditional dyes can be made and its reproduction are important for sustainable development.

6. Yogyakarta as a centre for higher education plays an essential role in the exchange of education and knowledge between the developed and developing world. Tourism is increasingly become a learning experience and the strength of batik lies in its ability to appeal to all levels of inquiry.

7. Information collection and documentation is vital to the collective history of batik as well as its continued development. The establishment of a database on batik, documenting its history, traditions, and designs would be a major educational, cultural, and industrial resource for Indonesia and the world.

8. Yogyakarta is the international centre for batik but it lacks a museum or heritage centre to communicate its place at the cutting edge of batik design and production. It is recommended that an innovative batik museum with the newest information, technology and with staff trained to the highest institutional standards be established in Yogyakarta.

9. It is recommended that a **Batik Development and Promotion Board** be considered to help in the promotion of the art and culture of batik throughout the world and in affirming Indonesia as the world centre for batik design, history, creativity and production.

Committee Members:

1. Prof. Dr Koesnadi Hardjasoemantri, SI (Chairman)
2. Prof. Dr Michael Hitchcock (Secretary)
3. Prof. Dr Moeljarto Tjokrowinoto (Member)
4. Pia Alisjahbana (Member)
5. Dr Kaye Crippen (Member)
6. Dr Puspitasari Wibisono (Member)
7. Amri Yahya (Member)
8. Dr Nasikun (Member)
9. Dr Sjafri Sairin (Member)

PART 1
CRAFTS, CULTURE AND TECHNOLOGICAL CHANGE

1 Quo Vadis Batik?

IWAN TIRTA

I have chosen the words 'quo vadis' especially because of the crisis situation which faces the traditional batiks of Indonesia today. Many of you probably remember the book and the subsequent movie by the name *Quo Vadis*, in which the apostle Petrus, fleeing the Imperial City of Rome from the atrocities inflicted by the Emperor Nero, encountered the Lord outside the city walls. In answer to his question 'quo vadis, Domine?', he was told to return to Rome and face the challenge.

Some parallels can be made with the crisis situation that confronts Indonesian traditional batik in the 1990s. Those who are concerned with the future and fate of traditional batik can ask the same question. Where do we go from here?

The alarming developments that have taken place since the early 1980s regarding traditional batiks but also other types of textiles in our Archipelago have made it necessary for us to re-examine or rethink the direction in which all traditional fabrics are going.

The situation is grave, but not severely so. The orange light is already flickering, so that we should be prepared when the red emergency light flashes on. I do not want to sound like an alarmist or over-pessimistic, but not enough people seem to be concerned.

Before continuing with this paper, it is important that we define again the words 'traditional Indonesian batik'. Traditional Indonesian batik is the process of surface textile decoration which uses wax as a resist. This particular resist cold dyeing technique involves the application of wax with both the *canting* as well as the *cap* or copper stamp. The main criterion is the use of wax as a resistant.

In former times batik *cap* was excluded from being regarded as 'true' batik, but by the 1990s it came to be regarded as hand *cap* batik.

When the *cap* method appeared in the late nineteenth century, purists were already sounding the alarm bell. Indeed the *cap* stamp was invented solely on the basis of commercial considerations. Not much artistry was involved in the process, although the making of the intricate copper wires on

3

a stamp required meticulous precision and some kind of artistry. The dyeing process, of course, even in *cap* batik, is still done by hand. Batik *cap* did not manage to eliminate real hand-drawn batik, but resulted in the erosion of hand-drawn batik.

In the year 1997 we were faced with another challenge. Both hand-drawn batik and batik *cap* were being pushed from the scene by other forms of textile decoration, in many cases wrongly using the label 'batik'.

What are the reasons for this new crisis situation? Several factors can be pointed out as the cause. First of all, there are the *internal factors*. By internal we mean that these factors lie within the batik process itself. Hand-drawn batik, as we all know, is a laborious procedure taking up long hours and days. With the increasing social demands and increasing living standards of both the batik labourers and entrepreneurs, wages and costs are also on the rise.

If formerly the bulk of the batikkers were living off the land, today they are increasingly landless and have become very urbanized. Wages are relatively high and entrepreneurs constantly try to increase their profit margins. It becomes more and more expensive just to pay for the labour costs.

The use of imported dyestuffs and imported cotton, although woven and spun in Indonesia, has also added to the production costs. Thus the cost of making superfine batiks and even batik *cap* has become an expensive undertaking.

If cost factors are pushing traditional batiks into the realm of expensive articles, another reason for their increased price is the *shortage of skilled batik workers*. Due to the low wage system that prevails, many of the batik workers are moving to other more lucrative jobs, mostly in the textile and cigarette-making industries.

The government's transmigration policy of moving groups of village families to other islands of the Archipelago has also depleted the number of batikers in Central and East Java.

As a consequence of the above-mentioned factors, batik entrepreneurs have started looking for cheaper ways of producing traditional batiks. Those who are more astute are making even more drastic short cuts to this laborious process by turning to screen-prints, both hand-screen as well as machine roller prints, borrowing freely from batik patterns and motifs. In doing so, these 'enterprising' batik manufacturers, wittingly or unwittingly, are assisting in the demise of traditional batik.

A second group of factors lies *outside the field* of the batik industry, namely within late twentieth century Indonesian society. One could point to the eclipse of the Javanese aristocracy or, in other words, the gradual disappearance of traditional arbiters of taste (this also applies to other ethnic groups in Java). There remain few traditional standards by which batik workers and manufacturers can test their excellence or quality. As a result there is a decline of design quality, the only criterion left is their commercial value or saleability in the big cities.

Since World War II, the number of people in Indonesia has increased by millions, and many of them, especially those who do not come from the island of Java, wear batik only because it is part of the prescribed national costume for women and not because they are familiar with its meaning or connection with their own cultural values. Notions about traditional batik have become blurred. Indonesians from other island besides Java wear batik, but they rarely have any idea of its meaning or origins. They treat batik almost like any other form of clothing, devoid of any meaning. Visual attraction is the main yardstick when buying batiks

The *decline of buying power* of the general public is also an important factor in the decrease of saleability of traditional batik, especially the hand-drawn variety, in recent years. If clothing was one of the primary needs of the average Indonesian, today other needs have replaced it in importance. Priorities have shifted.

The documentation or preservation of traditional batik is in its infancy. This phenomenon is not restricted to the world of batik, but prevails in other fields of traditional Indonesian textiles. Not enough research is devoted to reviving or documenting old batik patterns; publications by Indonesians are scarce. Therefore neither batik workers nor batik entrepreneurs can see the highlights or masterpieces of Indonesian traditional batiks.

The obstacles are plenty. Now the crucial questions have become: where do we go from here? How do we overcome those challenges? Can they really be overcome?

If we look back to the days when traditional batik was at its zenith, we find that a piece of hand-drawn batik was an exquisite way of expressing oneself in a piece of art. Batiks were exclusive, cherished and worn only on special occasions. They were ceremonial cloths embodying a whole gamut of values. Unlike their later development, traditional batiks were neither just saleable commodities nor items of daily wear.

Batiks were traded historically, but it was not until quite late in the colonial period that they were largely turned into a commodity. The question is now: can traditional batiks still be made according to age old laborious methods?

Who makes them now and in what format? Originally, traditional batiks on the island of Java were worn: *kain panjang* (long cloth), *sarung* (tubular cloth), *dodot* (ceremonial cloth), *selendang* (scarf or shawl) and *iket* (head cloth). As modernization and Western education spread, the above-mentioned ways of dressing became less popular. Today we see that the number of Indonesians wearing batiks in the traditional manner has diminished sharply. Only in the interior of Java do people wear batik *kain panjang* as daily wear: this goes especially for the men.

Javanese men of the late twentieth century wear the Western dress of trousers and jacket. Indonesian women also increasingly wear Western style clothes; only on certain occasions do they wear traditional batiks the old way (i.e. as sarongs or *kain panjang*). Advisors from abroad as well as in Indonesia stress the importance of diversification, although still using the traditional process of making batiks.

Actually the diversification process had already begun in the 1950s. Batiks were already used as table linens, men's shirts, and women's casual wear for the home, but only on a small scale. The conclusion was reached that in order to survive, the traditional batiks had to be diversified; not only its products but also its basic material. So far, traditional batiks have mostly been executed on cottons, in many cases imported from Europe or Japan.

Although greatly admired for its intricate patterns and its laborious process of making batik was not seen abroad as a luxury fabric because of its basic material, cotton. In order to become a prestigious fabric in international fashion circles, batik had to use a more luxurious fabric such as silk.

While collecting samples of batik throughout Java in the 1960s, Indonesian designers encountered places where in the past silk batiks were made, such as Juwana and Lasem on the north coast of central Java. These places made silk scarves for the inter-insular batik trade with destinations like Bali and Sumatera.

The silk batik trade was interrupted and subsequently terminated by World War II. Silk batiks became a rarity, available only in antique shops, mostly on the island of Bali.

In 1962, in order to find a way of reviving silk batiks, we started with the making of silk batiks. After making many mistakes, eventually the technique of making hand-drawn batik on silk was accomplished. The main problem in making silk batik was the removal of the wax. Today, silk batiks are sold in many boutiques and shops, using mainly silk habotais and crepe de chines or satins

In an attempt to diversify traditional batiks further, experiments were made with heavier weights of cottons, mainly for furnishing fabrics. Adapting the batik process to heavier weight cottons in longer yardage than the usual *kain panjang* or sarong also required a bit of experimentation.

Thus, we see that in the mid-1970s the diversification process of traditional batiks was well under way. There was a proliferation of batik shirts, dresses, curtains, upholstery material, tablecloths, napkins, bags, etc. A shot in the arm was given by the former Governor of Jakarta, Ali Sadikin, who decreed that long sleeved batik shirts for men were acceptable as formal wear at receptions.

In spite of the great enthusiasm that was the result of this diversification, there were still handicaps that could not be overcome, due to batik's inherent limitations when applied to industrial use. For instance, hotels and public buildings, which at first enthusiastically embraced batik as the Indonesian fabrics that could lend that 'Indonesian' touch, discovered to their dismay that it could not meet the demand for uniformity or consistency in colour when ordered in large quantities. As for colour and rub fastness, it is important to note that batik is a way of decorating surfaces of textiles in a cold dyeing process unlike the machine coloured textiles, where the yarns are boiled in hot reactive dyestuffs.

Another way of diversification was the emergence of batik 'paintings' or wall hangings. The art of making batik painting was already well established in both Singapore and Malaysia in the late 1950s. It 'hit' Java in the early 1970s, where it has remained a fad rather than becoming an important art movement, borrowing more from Western painting that traditional batik.

Some artists, like the late Sulardjo and his brother Sumihardjo and the famous batik painter Amri Yahya, have created striking works in strong colour with both abstract and realistic subjects, while the artist Ardianto used mainly faded pastel colours. Their tradition can now be found in the numerous workshops scattered around the City of Yogyakarta of which the Tamansari or Water Castle is the most well-known location. Although this 'modern' diversification helped the batik industry to find new ways of survival, it came at the cost of sacrificing the symbolic meaning of the patterns.

Due to the new uses of batik, motifs became more diverse, with the accent on more 'naturalistic' and less meaningful patterns (for the Western-oriented fashion world, which likes flowers). In order to appreciate the factors that play an important factor in the way the art of Indonesian batik is going, we must bear in mind that the direction can be twofold. There must be a constant source and evolution of creative ideas, producing design concepts according to standards, which are continually examined, scrutinized and refined. Otherwise, products degenerate into cheap junk.

On the other hand, with the commercial side, a more utilitarian product should be encouraged too, because it will provide a livelihood for millions of people. One cannot live from art alone. This side should be developed by private production coordinators, either Indonesian or foreigners, with the assistance of good product developers in Indonesia, as well as in foreign countries. They will constantly think of modern applications for traditional products and continually upgrade the quality, because doing so results in increased revenues. Until now, there have not been enough organized ways of developing this commercial side of handicraft. People just do it on their own. If they are lucky, they hit on the right product that is sophisticated and of sufficient quality for the market. Contacts are made privately with artists showing at government organized exhibitions, but such contacts are never properly publicized. The artistic side needs protection and constant nourishing from patrons and sources of inspiration like museums, libraries, exhibitions and workshops

The Indonesian craft-worker has little against which to measure his or her talent at the moment. There is a need for highly motivated, capable and discerning men and women to exercise control, criticize and direct quality. For people who care, who will expose crafts-people to the best from the past or what will become the future by showing them examples or photographs? The talents and the skills are there, and the costs are still relatively inexpensive.

Areas like Bali and Central Java are excellent examples of places where highly artistic products can still be produced. It is a long, hard battle to make people recognize the urgency for action of adequate documentation and preservation programme. Again two directions can be taken at the same time: one is to develop and preserve the artistic side; the other is to the develop the mass market utilitarian side without sacrificing either. Indonesia has come to the crossroads and must decide quickly. In the light of everything that has been mentioned, the time has come for Indonesians to decide in which direction the art of batik is going.

There are several textile associations who concern themselves with the fate of batik, but none has taken a definite step about preservation, documentation and development of batik. There is this confusion and profusion of products sold under the name of batik, but only as a 'flag of convenience'. Mostly these consists of screen prints, machine roller prints, using traditional batik patterns and motifs.

More and more Indonesians, especially the younger generation and inhabitants of other islands of the archipelago, have no clue or only a vague idea how and what the process of traditional handmade batiks is.

Quite a number of rather unscrupulous entrepreneurs are making use of the above fact: screen-printed batiks are made of the highest quality imitating traditional vegetable dyed batiks, even to the smell of beeswax. What solutions are to be undertaken when faced with this ever growing problem? Should people just write off handmade wax processed batiks?

For at least 20 years we have advocated the need for a 'batik mark', something like the internationally known 'wool mark'. Both the producers of genuine handmade wax-process batik as well as the customers should be protected from the flood of 'pseudo-batiks' masquerading as the genuine product. A batik mark can only be effective if it is backed by legal sanctions, penalties and legislation. Until this mark has been created the erosion and destruction of traditional handmade batik will continue. Today batik has already lost its third dimension and has become just another anonymous cloth. Only in a few cases does it still carry the contents of 'splendid symbols', and a 'voice' that speaks to us. Some scholars go even as far as saying that 'true batik' died after 1950.

We cannot turn the hands of the clock backwards, but we can slow down the destructive erosion that is taking place. Of the utmost importance in our fight against the disappearance of genuine batik is documentation by Indonesians as well as foreign scholars presentation of collections in museums scattered throughout Java. If Indonesians themselves do not take action, nobody will. Let us not be the cause of the loss of 'magic' in that fabled cloth called batik Indonesia.

2 Aspects of Intellectual Property and Textiles

BILL MORROW

My paper is structured around the two themes of 'tourism, batik and fashion' and 'constructing cultural identity through textiles', but before dealing with the intellectual property issues I have associated with those themes, it may be helpful to clarify what we mean when we speak of intellectual property.[1]

Consider the following three categories.

1 The International Intellectual Property Regime

This is the area familiar to most of us. It includes patents, copyright and trade marks and a number of other rights, including registered designs, plant varieties, circuit layouts and confidential information. Some would also include other areas of the law, such as laws protecting geographical indications of origin and special laws protecting symbols, for example flags and crests, but these are not generally regarded as property rights. There is also no unanimity about whether secret and confidential information is property. In this paper I wish to emphasize the distinction between rights which are rights of property (proprietary rights) and those which are not.

The international intellectual property regime is principally concerned with rights of private property. Copyright, patents, trade marks and registered designs are usually treated by national laws as property. Those laws create rights in intangible property. Issues relating to fair trading law and competition law impact upon the use of intellectual property rights.

Intellectual property rights are of course established by national law but these laws are regulated and influenced by the many multilateral and bilateral international conventions dealing with intellectual property. The existing international intellectual property regime is driven by the World Intellectual Property Organization (WIPO) and the World Trade Organisation Agreement on Trade-Related Intellectual Property Rights (TRIPS).

2   Indigenous Intellectual Property

This is an area of law not generally recognized by the existing international intellectual property regime, but it is an important aspect of the cultural identity of indigenous peoples. It has received international attention, through, among other things, the United Nations Draft Declaration of the Rights of the World's Indigenous Peoples, the UNESCO-WIPO Model Provisions for National Laws on the Protection of Folklore Against Illicit Exploitation and Other Prejudicial Actions and the Mataatue Declaration on Cultural and Intellectual Property Rights of Indigenous Peoples.

Indigenous intellectual property may involve many things such as rituals, dances, signs and symbols and know-how including biological and ecological know-how. There is a growing amount of discussion and writing internationally about bioprospecting and the rights of indigenous communities but that is an issue beyond the scope of this paper. In relation to textiles, however, indigenous intellectual property is of relevance to the use of traditional designs and traditional know-how about textile production.

Indigenous intellectual property is at times used to refer to things not treated as property either by systems of national law or by the laws of customs of indigenous communities. A distinction can and should also be made between indigenous intellectual and cultural property or rather cultural heritage.

3   Cultural Property and Cultural Heritage

The law when it deals with cultural property and cultural heritage usually does so by providing for the protection and preservation of works, things and places of cultural value or heritage. Such laws are ultimately concerned with the preservation of cultural identity and not with private proprietary rights. The term 'cultural property' is problematic as it raises questions about who is the owner and whether the rights are really proprietary in nature.

Tourism, Batik and Fashion

Under this heading I wish to consider the three most important areas of the present international intellectual property regime for batik producers, that is copyright, registered designs and trade marks. I will advance two propositions. First, that the protection offered to batik producers by copyright is not comprehensive and therefore copyright will not subsist in all batiks which are made. Second, that the protection offered by copyright is generally of more use to batik producers than a system of registered designs. I will refer to copyright law in Indonesia and Australia to support my arguments for those propositions.

Copyright

In recent years the origins of copyright law in the West have been critically re-examined. There has been interest in its connections with the thinking of the Enlightenment and speculation about whether this has resulted in an overly romantic and individualistic view of the author. There as been a view among some in Asia that such laws are relics of colonialism and, if not contrary to Asian traditions, then at least to Asian interests. It is within that context that the following comments are made.

The development of copyright law in the West was not uniform. There were differences, for example, between copyright in England which was viewed as being based on economic rights, that is, the right to exploit a creative work by controlling reproduction and publication, and the view in continental Europe that copyright was a personal right.

Indonesian law is based upon Dutch law, a civil system. Australian law is based on English law, a common law system. This difference explains why moral rights, most notably the rights of attribution and integrity, are available under Indonesian law (Article 24) but why they are still not available to artists and authors in Australia. It is expected that they will soon be available in Australia but this has been brought about largely as a result of a widely held perception that Australia was not complying with its obligations under Article 6b of the Berne Convention. Australia is a member of the Berne Convention. Indonesia was a member of the Berne Convention but withdrew in 1958. On 5 September 1997 Indonesia rejoined the Berne Convention but with a reservation regarding Article 33 (dispute resolution before the International Court of Justice).

To what extent do the copyright laws of Indonesia and Australia protect batik? In considering this issue it is worthwhile remembering that most batiks are not made by the person who created the design.

Indonesia In Indonesia copyright exists by virtue of the Copyright Law of 1982 which was amended in 1987 and 1991. The Copyright Law (Article 11(1)k) specifically protects batik or more particularly *seni batik* (batik art).

It is protected for the life of the creator plus 50 years (Article 26). Article 11 of the Copyright Law provides that all forms of plastic art (*bentuk seni rupa*) such as painting are protected, so presumably a drawing would also be protected.

Pursuant to the Copyright Law, a person who draws a batik design on paper will be the 'creator' (*pencipta*). The creator is a person or persons who as a result of their ideas, imagination, skills know-how and expertise, produces unique and personal pieces or work (Article 1.a). A creator can claim the rights given by Article 2 (publication and duplication).

What if a person other than the author of the design on paper takes it and applies it to a batik fabric? Under the Copyright Law that person cannot be a creator unless they interpreted the design creatively. Article 7, however, provides that the designer will be the creator even though the designer does not make the batik, but only if the batik is made under the designer's lead and supervision.

The use of the term *seni batik* in Article 11(1)f suggests that an artistic element is necessary, but presumably such an element is likely to be found in most batiks whether they are batik painting, batik *tulis* or batik *cap*. It is not clear however whether the artistic element must reside in the finished batik or in the skill of the person making the batik.

Under the Copyright Law where the designer also makes the batik, copyright is likely to subsist in the batik, it seems, however, that an artistic element is still required.

It appears, then, that the Copyright Law will not protect batiks which do not have an artistic component nor does it protect batiks which are not made under the lead and supervision of the designer (even though the batik itself may not be protected, copying that batik may still be an infringement of the original drawing).

The mere copying of traditional designs would not qualify a person under Indonesian Law as a creator because it would not involve any imagination and skill. Copyright may subsist if the person copying the work does so freely and imaginatively.

Batik designers and batik producers in Indonesia should be aware of Article 8(1) of the Copyright Law (1997) which provides that in the private sector the employee is the copyright owner unless otherwise agreed.

Australia The Australian Copyright Act, 1968 does not specifically refer to batik but copyright is available under the Act to 'artistic works' which by definition (Section 10) include drawing, painting, engraving and works of 'artistic craftsmanship'. It is likely that many batik works will qualify for protection in Australia but not all batik works will be protected. Those that do qualify for protection will be protected for the life of the author plus 50 years, if published during the author's lifetime (S33(2)).

In Australia, a person who draws a design on paper for a batik will be the author of that drawing. To qualify for protection under the Copyright Act, the drawing needs to be an original work and therefore if merely copied from another source such as a traditional design, it will not be original and will not qualify for copyright protection. If it is only loosely copied, then it will qualify for copyright protection to the extent that the copy is original. It is generally conceded that the threshold of originality is low.

If the person who designs the batik also makes it, then, under Australian copyright law, the batik work itself will qualify for protection as a 'work of artistic craftsmanship'. Other types of artistic works such as drawings and paintings are not required by the Australian Copyright Act to have any artistic merit. For a work to be a work of 'artistic craftsmanship', however, it must be 'artistic'. When the courts have had to consider this issues, they have had a great deal of difficulty in establishing what is meant by artistic craftsmanship.

Under Australian law the author of a work of artistic craftsmanship is the person whose skill and artistry produced the work. If it is designed by one person and made by another, the designer cannot claim to be the author of the work as a work of artistic craftsmanship.

In Australia there is a further gap in the protection offered to batiks. Batiks which have no artistic merit are probably unprotected unless the view is adopted that batik making by canting is a drawing, by a cap is an engraving and a batik made by brush is a painting. If that view was adopted, and there is no indication that it would be, then artistic merit would not be required. There does, however, remain the requirement of originality and if the person who makes the work is merely copying someone else's design then copyright will not subsist in the batik and, if made without the designer's permission, it will probably infringe copyright in the designer's drawings.

It is clear, then, that in Australia at least, a batik producer should retain all original drawings as there may well be copyright in the drawings but not in the finished batik itself. Where those drawings have been commissioned, the producer should consider whether it is necessary to acquire ownership of copyright in the design because the act of commissioning drawings will not necessarily vest ownership in the batik producer. The Copyright Act in Australia, unlike the Indonesian Copyright Law, provides that where the artistic work is made by a person pursuant to their employment, then their employer is the owner of copyright (S35(6)) subject to any agreement excluding or modifying that position.

Registered Designs

My second assertion is that copyright offers more attractive protection to textile designers than a system of registered designs. The chief advantages of copyright protection are that it is not dependent upon a system of registration and the period of protection is longer. Indonesian Law does provide for the registration of copyright, but this is not a prerequisite for protection (Article 29(4)). It does, however, create a rebuttable presumption that the person registered is the originator (Article 5). Australian Copyright Law has no system of registration. Australia, however, does have a design registration system established by the Designs Act, 1906 under which designs, including fabric designs, may be registered and obtain protection for up to 16 years (Section 27A). The benefits of registration include the right to take action against unregistered designs independently created. Under copyright law, two designers may independently conceive the same design, in which case copyright would subsist in each work and neither work would infringe copyright in the other.

It is interesting to note that the first English Act dealing with copyright in industrial designs was an Act of 1787 which gave to designers, printers and proprietors of new and original patterns for linens, cottons, calicos and muslins the sole right to print them for two months from the date of first publication. This was presumably meant to give them a flying start in the market. It should also be noted that it was not until 1862 that copyright protection was provided in England for drawings and paintings by the Fine Arts Copyright Act, 1862 (Fysh, 1974, pp. 2–5).

In Australia the Designs Act, 1906 grants exclusive monopoly rights to designs which are registered under that Act. The broad intention has been to exclude industrial designs from copyright protection when those designs are industrially applied. Therefore if a design is industrially applied it must be registered under the Designs Act if it is to receive any protection. It has also been intended that designs which are principally artistic in character should receive the protection afforded by copyright. There has however been a great deal of difficulty in Australia clarifying the overlap between design and copyright protection.

As the law presently stands in Australia, a batik designer who produces a 'new or original design' on paper no longer loses copyright protection if it is industrially applied, whether it is registered as a design under the Designs Act or not. The designer therefore can retain copyright protection and register it as a design. If, however, the work is one of artistic craftsmanship then such dual protection is not possible. In that case, if the design is industrially applied, copyright protection is not lost; however, copyright protection is lost if the work is registered as a design (S75). These provisions make it all the more important for batik designers to retain their drawings and paintings of designs in order to rely upon those as the basis for registration or to support a claim of copyright ownership.

The design system has not proved popular for textile designers in Australia. This is reflected in figures generously provided by the Australian Industrial Property Organisation for this paper, which reveal that in 1996 only 51 designs were registered for textiles (44 in 1995 and 56 in 1994). This may be partly because of a perception that design protection is limited ('easy to get and easy to get around'). In addition, most fabric designers, particularly in the fashion industry, see their designs as having a short commercial life-span which does not warrant the trouble and expense of registering a design. This is not to be totally disparaging about a system of registered designs for textile designs. Some designs do have an extended life-span and may be worth registering. The Australian designer Ken Done, for example, has registered designs for a number of textiles. A commercial textile producer may find the registration of certain designs strategically useful.

Deciding upon the best manner of protecting textile designs is a problem facing a number of countries. Article 25(2) of the TRIPS agreement provided:

> Each Member shall ensure that requirements for securing protection for textile designs, in particular in regard to any cost, examination or publication, do not unreasonably impair the opportunity to seek and obtain such protection. Members shall be free to meet this obligation through industrial design law or through copyright law.

Indonesia does not presently have laws protecting industrial design although it appears that laws are being prepared (Kesowo, 1996).

Trade Marks

Most of what I have said so far is of relevance mainly to batik designers and producers. Tourists usually have little interest or knowledge of the copyright ownership of textile designs. A trade mark or trade name, however, may be of more significance to tourists. After the issue of price, the concerns of tourists may include some or all of the following:

- whether the textile is handmade or not (e.g. *tulis*, *cap* or silk screen);
- whether the dyes used are traditional natural dyes or synthetic dyes;
- whether the textile was made in the place from which the design originates;
- the quality of the materials used;
- the working conditions of the people making the textiles.

In the absence of specific information, the trade mark or trade name may be sufficient reassurance to tourists with respect to some or all of these issues.

Constructing Cultural Identity Through Textiles

There is no doubt that textiles play an important part in creating and reflecting the cultural identity of many cultures throughout the world. This is clearly the case in Indonesia where the country's rich textile traditions create identities for the many diverse cultures spread across the Indonesian Archipelago.

In this section of the paper I want to consider several issues, including what rights, if any, indigenous communities have, or should have, in relation to their traditional textile designs and whether, or to what extent, the appropriation of those designs by other cultures should be tolerated.

A characteristic of the culture of the Western world in the second half of the twentieth century has been its tendency towards cultural appropriation. The advent of tourism and the mass media have permitted easy access to other cultures. Javanese culture, for example, is beginning to appear on the Internet (Field Museum) and in recent years there has been a rash of books published on batik and batik designs.

Much of the cultural production of indigenous people is available for appropriation. Textile designers commonly appropriate imagery and styles from other cultures. Many tourists engage in cultural appropriation whether it be as photographers or collectors of cultural objects. We appropriate aspects of other cultures not only to increase our understanding of those cultures but also to construct an identity for ourselves in the post-modern world. Appropriation is often good, but not always.

An Aboriginal artist who almost gave up painting a decade ago after he discovered his work had been copied by a fabric designer, won the country's most prestigious award for indigenous art yesterday (*The Australian*, 16–17 August, 1997, referring to the artist Yanggarriny Wunungmurra).

Appropriation from some cultures is deeply offensive to the people of those cultures. Appropriation may also be an infringement on traditional proprietary rights (native title).

In Australia many aboriginal groups have a recognisable form of indigenous intellectual property. Strehlow (1965) pointed out that '… all things connected with the old religion of Central Australia – myths, songs, dramatic acts, and sacred objects – were privately owned'. The disclosure of information or the improper exercise of 'rights', for example, copying designs without the permission of the owner or custodian, attracted severe penalties within those communities. Other communities also developed rights in intangible property. Ruth Landes (1937, p. 109), noted that 'Certain dances of the Ojibwa are commodities, in that they are bought and sold'.

A number of societies have dealt with rights in intangibles as proprietary rights. Not all of those rights can be easily recognised as rights or private property under the present international intellectual property regime. However, many of the rights are analogous to rights of property and where those rights are still observed by indigenous communities they are capable of recognition by the wider community under national laws. Such recognition does not require special legislation protecting a generalised right for all indigenous people within a nation nor does it represent a challenge to the sovereignty of the state. It involves the recognition of the proprietary rights which exist in indigenous communities (native title). It also requires recognition of the diversity of those rights within nations.

As I have already pointed out, copyright does not protect traditional designs. The life of copyright is limited and traditional designs form part of the public domain and are available for use by anyone. A designer in Europe, for example, can copy a traditional batik design and apply it to a silk screen dress fabric. Similarly, an entrepreneur in Singapore can apply traditional batik designs to wrapping paper.

Australia does not have any specific protection for indigenous intellectual property. There is, however, a great deal of state and federal legislation dealing with Aboriginal sites and objects. It was, in fact, only in 1992 in the landmark High court decision in Mabo (175 CLR 1) that the doctrine of terra nullius was rejected and Aboriginal traditional rights in land recognized. Aboriginal people, however, have had difficulty protecting traditional designs. A number of copyright and confidential information cases have been brought by aboriginal people in Australia. I wish to mention only one of those cases, namely, the 'carpets case' as it is known (Milpurrurru v. Indofurn Pty Ltd (1994) 30 IPR 209). That case involved the importation into Australia of carpets made in Vietnam using artwork by aboriginal artists based on traditional aboriginal designs. The aboriginal artists had not consented to the reproduction of their artwork. What is notable about this case is the sensitivity shown by the court to cultural issues albeit within the confines of copyright law, in particular the works were found to be original works even though based on traditional designs and additional damages were awarded against the infringers to reflect culturally-based harm. Five of the aboriginal artists were dead and in accordance with aboriginal custom their names were not spoken during the trial.

This case is indicative of the growing awareness within Australia of the traditional rights of Aboriginal and Torres Strait Island people and of the short-comings of existing Australian intellectual property law in protecting their rights. There are calls for the protection of folklore and proposals are being advanced for the introduction of authentication and certification trade marks to identify works which are made by Aboriginal Torres Strait Islander people.

Indonesia attempts to protect aspects of its diverse cultural traditions through a form of state ownership. Article 10 of the Indonesian Copyright Law provides that:

(1) The State shall hold the Copyright to works of archaeological and historical remains and other objects of national and cultural significance.

(2) a. Publicly owned works, such as stories, legends, fairy tales, folktales, epics, songs, handicrafts, classical and folk dances, choreography, calligraphy and other works of cultural and artistic significance, shall be maintained and protected by the State:

b. Copyright to article (2)a. is held by the State and is applicable outside the country.

(3) State Copyright of works described in this article are further stipulated in the Government Regulation.

The copyright referred to in Article 10(1) appears to be perpetual. It is also apparent that all traditional batik designs, including the Sultan's 'forbidden patterns', are included under Article 10. Can they only be reproduced with the permission of the state or are all traditional batik designs the common possession of the citizens of Indonesia? Can an Indonesian citizen use the traditional designs of any Indonesian culture, for example, the use of Sumba designs in the weavings produced in Japara, Java, or the use by a Javanese batik maker of Batak or Toraja designs?

It is interesting to speculate about whether Indonesian law should go further. For example, should it provide that batik produced with a Cirebon design should indicate whether or not the batik was produced in the Cirebon region in the way that TRIPS deals with the geographical indications of origin for wine and other products (Articles 22 and 23)? An obvious problem with such an approach is that people move about, both voluntarily or involuntarily, and carry their traditions with them. There is also a danger that we may become overly protective towards cultures and fail to recognize that in some regions there are traditions of appropriation. For example, many Cirebon designs have been created by appropriating imagery from a wide number of cultures including China, India, the Middle East and Europe and yet in the Pasisir there has also been a good deal of blatant copying of other people's work including the copying of fairy tale book illustrations (Veldhuisen, 1993, p. 41).

Should the Sultan of Yogyakarta be entitled to object when his crest or any of the 'forbidden patterns' such as Parand Rusak, Sembagen Huk or Garuda Ageng (Boow, 1988, p. 70) are used? Are the Sultan's traditional rights to the patterns proprietary in nature or are they based upon rights of sovereignty which have now been extinguished? The latter would appear to be the case.

> Batik patterns were important in the codes of dress and regalia in the Courts of the old State of Mataram. Motive and design styles indicated status and political allegiance. They showed ceremonial and political precedence and there were rights to particular patterns. Conformity to customary dress law showed homage. Disrespect for these laws was an act of rebellion (ibid., p. 14).

An interesting comparison is the use (and restriction on use) by the British monarchy of the royal arms and the use of armorial bearings (Halsbury, 1984, pp. 224–5). In 1954 for the first time in 200 years the Court of Chivalry was convened to hear a petition from the Manchester Corporation asking for such reparation as the Court should think fit against the defendant, the Manchester Palace of Varieties Ltd ([1955] 1 All ER 387. The defendant had displayed the Corporations' arms in its theatre auditorium and used the arms on their common seal. Lord Goddard, CJ, found that 'The right to bear arms is, in my opinion, to be regarded as a dignity and not as property within the true sense of that term'.

Lord Goddard made an order prohibiting the use by the defendant of the

Corporation's arms but indicated that had the defendant only displayed the arms it may not have been objectionable. He said:

> I am by no means satisfied that nowadays it would be right for this Court to be put in motion merely because some arms, whether of a corporation or of a family, have been displayed by way of decoration or embellishment. Whatever may have been the case 250 years ago, one must, I think, take into account practices and usages which have for so many years prevailed without any interference. It is common knowledge that armorial bearings are widely used as a decoration or embellishment without complaint (p. 394).

I am told that in the 1990s in Yogyakarta the Sultan's patterns are not copied out of respect for the Sultan. It appears, however, that the eighteenth century decrees of the sultans of Yogyakarta and Surakarta concerning batik patterns were not respected in the Pasisir (McCabe Elliot, 1985, p. 68).

No culture is static and at the end of the twentieth century few cultures are isolated. Nowhere is this more apparent than here in Yogyakarta where the culture is an overlay of Islamic, Hindu, Buddhist and indigenous traditions with aspects of Chinese and Western culture. The culture here is described as syncretic because of its ability to reconcile different systems of belief.

But even robust cultures are vulnerable. Indonesian textile traditions such as batik *luri* are beginning to disappear (Kakiailatu, 1996) and will need special protection or patronage if they are to survive. Old traditions such as *wayang kulit* are increasingly seen by the urban young of Java as boring and irrelevant. Asian countries are bombarded with B-grade Western films, commercials and attitudes.

Given these pressures it is understandable that many countries should want to protect the traditions that give these countries and cultures their identity and to object to the appropriation of cultural traditions by outsiders. The issues involved are not easy ones to resolve particularly in view of the fact that many countries, including Indonesia, do not have a single cultural identity and appropriation by one cultural group from another cultural group within the country may be as objectionable as appropriation from outside the country.

There are often good reasons for preventing or restricting appropriation but against those we should weigh the positive aspects of appropriation such as the development of an understanding of other cultures and the enrichment of our own cultures.

Note

1   I first came to Yogya in 1977 with my wife, Colleen Morrow, to undertake a short course in batik making conducted by the Batik Research Institute. It was the beginning of what has been for both of us a long and rewarding association with Indonesia and Indonesian textiles.

References

Boow, J. (1988), *Symbol and Status in Javanese Batik*, Monograph Series No. 7, Asian Studies Centre, University of Western Australia.

Elliot, I. McCabe (1985), *Batik: Fabled Cloth of Java*, Harmondsworth, Viking.

Field Museum of Natural History, Chicago, http://www.fmnh.org/exhibits/javamask/ Javamask.htm (accessed 1 September 1997).

Fysh, M. (1974), *Russell-Clarke on Copyright in Industrial Design*, London, Sweet & Maxwell.

Halsbury's Laws of England (1984), 4th edn, London, Butterworth.

Kakiailatu, T. (1996), 'Luik Tenun Gendong: Yesterday's Relic?', International Symposium on Indonesian Textiles, Jambi.

Kesowo, B. (1996), 'Indonesia's Profile', APEC Industrial Property Rights Symposium, Tokyo, http://www.jpp-miti.go.jp/pate/repo/apec/idone.htm (accessed 24 September 1997).

Landes, R. (1937), 'Ojibwa Sociology', *Columbia University Contributions of Anthropology*, Vol. 29.

Strehlow, T.G.H. (1965), letter to the Chairman of the Select Committee on the Native and Historical Objects and Area Preservation Ordinance, published in The Legislative Council for the Northern Territory (1965), *Report from the Select Committee on the Native and Historical Objects and Areas Preservation Ordinance 1955–1960*, Darwin, Commonwealth Government Printer, p. 8 (Appendix II).

Veldhuisen, H.C. (1993), *Batik Belanda 1840–1940: Dutch Influence in Batik from Java. History and Stories*, Jakarta, Gaya Favorit Press.

Acknowledgement

The author acknowledges the kind assistance of Mr Heri Herjandono of the legal firm Iman Sjahputra & Associates in answering his many questions about Indonesian copyright law.

3 Innovation, Change and Tradition in the Batik Industry

TERUO SEKIMOTO

This chapter does not deal with the 'art and craft' aspect of Indonesian batik. Instead, it first examines the way batik making in Java has adjusted itself to the ever-changing circumstances of modern society and developed as a form of an economic enterprise since the last century. A brief sketch of the development of batik industry in Java will demonstrate how a contemporary craft, with its rich heritage from pre-modern times, retains traditional elements under both favourable and difficult conditions imposed by the modernity. In order to think about tradition in a meaningful manner, I assume that the tradition is not a product of the past but a particular conceptual framework by which we see things around us. It does not belong to the past but to the contemporary experience of ours.

Secondly, this chapter briefly deals with the problem of what I call the 'traditionalist' and the 'modernist' view surrounding batik. The traditionalists emphasize the protection and preservation of the past, and are against any changes and development. Opposing them are the modernists who cast a cynical look at 'dying' old things. Some of the 'invention of tradition' theorists such as Hobsbawm and Ranger are a more sophisticated version of them. What we learn from the modern history of batik will be able to correct the mistake of the dichotomy between the old and the new, the tradition and the modernity. A mediated, dialogical view thus obtained will be quite relevant to the general problem of how local cultural identities find their expression in an era of globalization.

Batik as a Cultural Artifact

There is in Japan a small but persistent demand for Indonesian batik which is met by importers and retailers not only in large urban centres but small, local cities. Even as early as in the 1920s, there was a Japanese firm, Fuji Yoko in

Yogyakarta, making broad bands (obi) of batik for traditional Japanese clothes and other batik products, and exporting them to Japan (Kat Angelino, 1930–31, pp. 175, 189). The Japanese import of batik has been small in quantity compared with major commodities Japan imports from Indonesia such as crude oil, LNG, plywood, fish, textile, or garments. However, batik has its own characteristics which is different from these major imports. Whereas the consumers of the common types of commodities do not concern themselves about the country of origin, batik is recognized as a distinctively Indonesian product. Unlike garments or plywood, countries other than Indonesia (and Malaysia) do not enter into batik production even if there is a sizable market for this particular commodity. Commodities are often not closely associated with their origins, but this economic dictum does not apply to batik.

Batik is produced mostly in Indonesia, especially in Java. In the eyes of those who appreciate batik, this association would appear to be straightforward, since batik is deeply rooted in the history of Java and Indonesia. Already in the sixteenth century European travellers reported the production of colourfully painted cloth in Java (Elliott, 1984, p. 36). One might maintain that batik's value lies in its links with the old traditions of Java, and therefore it cannot be transplanted to other countries. One would argue further that batik, in essence, is not a commodity but a cultural artifact.

Such a view of batik, which is based on a dichotomy between economy and culture, is not completely wrong. In 1990s Japan, batik is seen mainly as a cultural object. Articles dealing with batik would appear not in economic but cultural sections of Japanese newspapers. We should, however, note that this view of taking batik exclusively as a cultural object is forgetful of the modern history of batik industry since the last century, in which the dichotomy between the economy and the culture has not been self-evident. For example, there was the case of batik production in Japan and its export to Indonesia in the 1930s (Japan Textile Design Center, 1960, p. 19).

The Development of the Batik Industry Since the Nineteenth Century

There are only a few historical sources concerning batik production before the nineteenth century when it began developing as a modern business activity. Little is known about batik produced for sale in markets. Available sources indicate that Javanese courts in the seventeenth and eighteenth centuries had many women at their command, having them make hand-drawn batik for princes, courtiers, and their families. The sources also show us that a large amount of fine cloth – both woven and dyed stuff – was imported from India and used by local elites for ceremonial purposes. The Indian textiles, which dominated the world market before the industrial revolution in Western Europe, much influenced batik and other textiles in Indonesia. On the other hand, Indian manufacturers often intentionally modelled their products after the local textile designs of Indonesia in order to enhance sales there (Yoshimoto, 1996).

When European cotton print began penetrating into Indonesian markets early in the nineteenth century, its chief rival might not have been Javanese batik but Indian textiles. Without the development of the stamp method and consequently of the batik industry in Java, European cotton print might possibly have overwhelmed the domestic production of textile in Indonesia. In 1892, several Dutch residents in Java conveyed their apprehension that European imitation batik could eventually destroy the local production of batik (Koperberg, 1922, p. 147). Their foreboding proved wrong as the batik industry continued growing.

Produced at large textile factories in Japan, imitation Javanese batik was aimed to meet the demand for an inexpensive kind of batik among Indonesian consumers. In the early decades of the twentieth century, the use of Western-style pants or skirts was still limited in Indonesia. Batik wraparound skirts, which Javanese elite classes used to wear exclusively for centuries, also began to be adopted more widely by commoners, as a cash economy and wage labour made inroads into Javanese cities and countryside in the latter part of the nineteenth century (Shiraishi, 1990). Since there was a large demand for batik in Java and nearby areas, first British, then Dutch textile manufacturers began making imitation batik by using the newly-invented printing method and exporting it to Indonesia in the early nineteenth century. The Japanese manufacturers were only latecomers in the competition for the market. These efforts to export batik to Java, however, eventually failed as foreign manufacturers lost to competition from Javanese batik makers. Whether European or Japanese, imitation batiks failed primarily because their products were not colourfast (Brenner, 1991, p. 37). With the printing method in Europe and Japan available then, imitation batik was ultimately no match for Javanese products in which the time-consuming wax-resist method was applied on both sides of cloth.

At the time when Japanese manufacturers tried to export batik to Indonesia, Javanese batik makers made an inexpensive type of batik by the *cap*, or stamp method. In an older method of hand drawing or *tulis*, melted wax is filled into a copper tool or *canting*. A female worker holds the tool by hand, applying wax on cloth similar to drawing a picture. It is a highly time-consuming method. The stamp method, which appeared in the northern coast area of Java in the 1840s (Rouffaer and Juynboll, 1914), eventually revolutionized batik making. With this new technique, a far greater speed than the hand-drawing method became possible. As many minor improvements were added to the stamp over years, it gradually developed into a more sophisticated, larger tool, increasing the productivity of batik. Now batik making was segmented into two divisions. The hand-drawing method continued to produce quality batik in a smaller quantity while the stamp method turned out inexpensive batik in a much larger quantity.

Together with the new stamp method, new batik firms grew throughout Java and parts of Sumatra, each employing tens, or even hundreds of workers. The technical innovation made possible the mass production of batik at much cheaper cost, which then created new demand, further increasing the scale of the production and the number of firms. This was the typical case of economic development triggered by a technological innovation, even though the stamp method was still totally dependent on manual labour. The stamp method thus opened up new opportunities for modern forms of business endeavour to the non-European population in Indonesia, especially the Javanese. Since major industrial sectors in Netherlands Indies then were dominated by Europeans, batik making, along with the manufacturing of silverware and clove cigarettes, provided one of precious few niches in which Javanese entrepreneurs could prosper.

The productivity of batik further increased early in this century, as chemical dyes from Britain and Germany gradually replaced natural dyes of the traditional sort, thus, enabling a quicker dyeing process. Together with the stamp method and chemical dyes, the batik industry was prospering all over Java, even though short-term ups and downs were an inevitable part of this indigenous industry. In the 1920s, the owners of thriving batik firms built luxurious houses adorned with imported marble, chandeliers and automobiles. Timeworn buildings in the cities of Solo and Yogyakarta survive as testimonies to the past glory of the industry.

Because of its old tradition, one tends to romanticize batik, often seeing it as an antithesis of modern capitalism. The fact is, however, that batik has played an important role in the modern economic development of Indonesia. The tradition of batik, as we see it today, is as much a result of competition and technological innovation since the latter part of the nineteenth century as it is the continuation of pre-modern, indigenous tradition. The aforementioned episode of Japanese-made batik happened in the process of modern economic development in which a technological innovation engendered new demands and also an additional increase in production. In the 1990s the technology of printing in modern textile industry, which in the early twentieth century was not competitive enough to win markets from the local batik industry, is much more advanced. The latest laser print technique can produce an exact copy of the finest hand-drawn batik. Even the most knowledgeable experts of batik would have difficulty distinguishing between the original and the copy. However, such an idea sounds somewhat ridiculous since the commercial value of batik in these days is inseparable from the values attached to such notions as traditional art, handicraft, and Indonesian cultural heritage. Batik in the early twentieth century, on the other hand, was for the most part a common type of textile good whose value was judged simply by its price, use value and durability. The production of batik in Japanese factories was not in itself a strange act so long as it was cheap. The Japanese-made batik eventually lost competition primarily because of its technical shortcomings, namely, the fast deterioration of colour.

The World Economy and the Modern Tradition of Batik

It was the stamp method that made the modern development of the batik industry possible. One could, however, argue that whatever innovations might have been added to batik making since the last century, the art of hand-drawn batik which preceded them remained significant. Although detailed description is lacking concerning the 'hand-painted cloth made by Javanese women' mentioned by European travellers in the seventeenth and eighteenth centuries, batik at the beginning of the nineteenth century was not vastly different in terms of colours and motifs from today's 'traditional' batik (cf. the colour lithographs of batik-clad men and women in Raffles's *The History of Java*). In the field of textile and fashion design, batik is known internationally. This would not be possible without the aura of age-old tradition about it. Even if the modern batik industry is by and large an economic activity, the art of hand-drawn batik may well represent Java's cultural heritage. However, it, too, has undergone changes and developments in the modern social environment. It is in this social context that traditionalist discourse on batik has developed.

As is mentioned above, batik making was diversified into two sections after the invention of stamp method; expensive hand-drawn batik on one hand, and cheaper stamp batik on the other. In most cases, however, a batik firm produced both types of batiks within a single workshop and with a sexual division of labour; men worked on stamp batik while women made hand-drawn ones. Only a little is known about the origin of stamp batik, but according to one theory, stamp was invented not to facilitate mass production but to obtain finer, more exact patterns than hand-drawn batik could accomplish. The development of the stamp technique, then, further stimulated the development of hand-drawn batik into finer, more exact patterns.

The different types of hand-drawn batik are ranked by price and quality. Among hand-drawn batiks produced today, those of higher ranks are characterized by the machine-like precision in producing fine patterns. This particular attention to small details may well be regarded as an example of what Clifford Geertz (1963) calls the 'involution' of culture in colonial Java. The increasing precision of patterns occurred not only in hand-drawn batik but also in stamp batik. A new technique, which was introduced in 1930, made it possible to remove even the smallest gaps between stamped wax patterns so that uninformed laymen could no longer distinguish hand-drawn batik from stamped ones (Veldhuisen, 1993, p. 59). The time-consuming needed work to obtain the extreme precision of patterns is one of the reasons batik is much praised. It is, however, not a residue of old tradition preserved statically but a result of an accumulated process in which the modern business of batik making has developed over years.

The industrial revolution in Europe also influenced the modern development of batik. The tradition of batik, which we find today, has not been an isolated, local phenomenon but tightly entangled with the world economy. Up to the end of eighteenth century, batik makers utilized either domestic cotton or Indian imports: the former was thick and coarse, while the latter finer and highly valued. The first shipments of factory-made cotton from Britain reached Indonesian markets in the early nineteenth century. Then, Dutch cambric began entering, and gradually dominated the market. It was a type of cotton cloth woven densely of fine threads effecting the tight and smooth surface of textile. With the use of the cambric, more intricate patterning of batik became possible than it had been (Elliott, 1984, p. 38). The Netherlands, with its newly-emerging textile industry in Twente, found in Indonesia a very promising export market. As they were not successful in their effort to export print cloth, they eventually concentrated on exporting white cotton as a raw material for Java's batik industry (Van de Kraan, 1996). Thus, the development of the latter was helped partly by an export promotion policy sponsored jointly by the Dutch colonial administration and textile industry in the metropolis. From the 1910s on, however, Japanese cotton exports began making inroads upon Java and entered into harsh competition with Dutch cotton, especially in the 1930s (Saraso, 1954).

It was not just through the imports of white cotton, chemical dyes and paraffin that batik was tied to outer world. Its designs also underwent a significant new development under the influence of foreign textile designs. Since cloth was one of the major commodities in long distance trade long before the rise of modern capitalist economy, batik was never isolated from foreign influence. As is mentioned before, Indian influence on batik and other Indonesian textiles were especially strong during the seventeenth and eighteenth century. The stamp method and the subsequent rise of modern batik industry, however, brought about an overall new pattern in the production of batik. Besides the older centres of batik making in inland court cities, new centres were booming in some port cities on the north coast of Java. While manufacturers in the court cities such as Solo and Yogyakarta largely maintained the older batik designs of court circles, their rivals in the port cities adopted foreign designs much more freely. Brighter colours and motifs reflecting Chinese, Indian, Persian, and Arab influence, as well as ornamental designs from modern Europe, were incorporated and adapted to the existing batik tradition. Batik manufacturers there were ethnically more mixed, including Chinese, Arabs, and Eurasians side-by-side with Javanese. At the time manufacturers in the court cities were mainly ethnic Javanese, largely concentrating on the original kind of batik products, that is, square cloth for Javanese traditional wraparound skirts and shawls. Products were more diversified in port cities in the north, suitable for the then-emerging lifestyle of urban middle class. Now batik was designed, too, to be suitable for making Chinese and modern European dress, interior decorations, and so on. What developed in the port cities was a type of modern business activity which always sought new demand by creating one after another of newer designs and fashions.

Tradition as a Part of Contemporary Reality

Today's batik represents a heritage of fine handiwork and artistry. This tradition is generally regarded as essentially Javanese and as the antithesis of modernity. According to this view, which I would call the traditionalist view, the golden age of batik lies in ancient times and every change the modern era has brought to batik is a negative one: modernity always means decay in the tradition. As a consequence, batik becomes something which must be protected from the change of time. As we have seen, however, today's batik owes a significant part of its fame to innovations and changes since the last century. And it is exactly this modern development of batik that has given birth to the traditionalist view of it. This view first gained strength among Javanese and Dutch elite circles at the turn of the century (for a typical traditionalist view on batik by a Dutch writer, see Kats, 1922) and since then has been reproduced countless times up to now. The same concern about the decline and possible extinction of the batik tradition has been repeated for more than a century. This enduring sense of anxiety has its grounds, since the batik industry has never been a dominant force in modern economy. It has had to overcome one difficulty after another, and still has to. It should be noted, however, that batik making has survived and developed as a form of modern industry representing Indonesian traditions. The very fact that the same concern has been repeated over a century rather paradoxically demonstrates batik's potential for long-term resilience. If batik traditions had been dead, the traditionalist concern about batik, too, would have been extinct much earlier.

The traditionalist view of batik regards tradition and modernity as mutually incompatible, but modernity is a necessary precondition on which consciousness of the tradition of batik is formed. Modernity, however, is not an absolute condition in our contemporary life in which many different layers of modernity and tradition exist side-by-side in continual conflicts and contradictions. Jim Supangkat, an Indonesian art critic, writes that 'the actual traditional culture has been marginalized, not only by high art, or Western art, modern art and international art, but also by the (locally formed) concept of "traditional culture" itself' (Supangkat, 1997, p. 84). In other words, the reification of culture and tradition by the official discourse only marginalizes the actual manifestation of the tradition. 'Tradition' or 'traditional culture' are meaningful to us not because the past has a special privilege in itself. They are meaningful because both tradition and modernity are living realities of our contemporary life.

References

Brenner, S.A. (1991), 'Domesticating the Market: History, Culture, and Economy in a Javanese Merchant Community', PhD dissertation, Cornell University.
Elliott, I. McCabe (1984), *Batik: Fabled Cloth of Java*, New York, Clarkson N. Potter.
Geertz, C. (1963), *Agricultural Involution: The Process of Ecological Change in Indonesia*, Berkeley and Los Angeles, The University of California Press.

Hobsbawm, E. and Ranger, T. (eds) (1983), *The Invention of Tradition*, Cambridge, Cambridge University Press.

Japan Textile Design Center (1960), *Batikku: Jawa Sarasa no Moyo (Javanese Batik Motifs)*, Tokyo.

Kat Angelino, P. de (1930–31), *Batikrapport (Deel II, Midden-Java)*, Weltevreden, Kantoor van Arbeid.

Kats, J. (1922), 'De achteruitgang van de batikkunst,' *Djawa*, 2, pp. 92–5.

Koperberg, S. (1922), 'De Javaansche batikindustrie', *Djawa*, 2, pp. 147–56.

Raffles, T.S. (1978 [1817]), *The History of Java*, Kuala Lumpur, Oxford University Press.

Rouffaer, G. and Juynboll, H. (1914), *De Batikkunst in Nederlandsch Indie*, Utrecht, Oosthoek.

Shiraishi, T. (1990), *An Age in Motion: Popular Radicalism in Java, 1912–1926*, Ithaca, Cornell University Press.

Supangkat, J. (1997), *Indonesian Modern Art and Beyond*, Jakarta, The Indonesian Fine Arts Foundation.

Van Den Kraan, A. (1996), 'Anglo-Dutch rivalry in the Java cotton trade, 1811–30', *Indonesia Circle*, 68, pp. 35–64.

Veldhuisen, H.C. (1993), *Batik Belanda 1840–1940: Dutch Influence in Batik from Java*, History and Stories, Jakarta, Gaya Favorit Press.

Wirodihardjo, S. (1954), *Ko-Operasi dan Masalah Batik*, Jakarta, Gabungan Ko-Operasi Batik Indonesia.

Yoshimoto, S. (ed.) (1996), *Shirarezaru Indo Sarasa (Indian Textiles in Indonesia)*, Kyoto, Kyoto Shoin.

PART 2
TRADITIONAL BATIK

4 Cultural Values and Traditional Batik Patterns

OETARI K.W. SISWOMIHARDJO

Batik in Indonesia

Since the 1970s, batik in Indonesia has undergone marked change. Based on my findings in several places in central Java and mainly in Solo, the traditional use of batik appears to have declined.

Formerly, batik was a necessary item in people's daily life. Though its chief function was as part of the daily dress, it had been of use in countless other ways, beginning the moment a baby was born until the day s/he died. Batik is not only beautiful but it is also meaningful. Time changes and so do the functions of batik. Some of the former functions have been replaced, mainly in the drive towards efficiency. In modern life, for instance, we very seldom see the traditional *kain* and *kebaya* in public places like streets, stations or public transportations except in traditional market places. Even in the country where countrywomen used to wear *kain* and *kebaya* in a more casual way than the townswomen, this way of dressing has become rare. On the other hand, new usages of batik have been invented in a development process that appears to be unlimited.

Many factors have stimulated the ongoing popularity of batik: the change in batik material; the more colourful appearance; and the big role of designers in modifying the wearing of batik as a cloth. The Indonesian habit of wearing uniforms for events and ceremonies certainly adds a lot to the popularity of batik. Although batik has been generally accepted and admired, it is more than simply a visually pleasing garment or accessory.

Some evidence of an alternative way of looking at batik can be seen in traditional markets and shops, and even the larger department stores, and piles of batik products in all kinds of forms are shown and sold out. But the customers base their choice on the fabric, colour, design, price and the belief that some patterns are wanted for some special events, though often without being able to name the patterns. Very often this ignorance is also found in the sellers.

This is very much to be regretted, since batik's beauty is based on its visual and spiritual properties.

The Spiritual Part of Batik

Many Indonesians still subscribe to the strong belief that certain special batik patterns are needed for some important events in their life, like birth, marriage and death. An ongoing custom indicates the strength of these associations. When someone dies, the mortal remains are covered with a batik cloth before it is put in the coffin. However, many of the younger generation do not know the names of those patterns, let alone their spiritual content.

It is the spiritual part that makes batik so special among other designed textiles. Traditional batik patterns are not only beautiful, but they also contain symbols of values and customs, too precious to be forgotten. In those patterns can be found ornaments which have the form of geometrical figures, fauna and flora and parts of the universe. It is those symbols together that comprise the spiritual part of a batik pattern.

The geometrical figures consist of dots, lines, circles, rectangles and so on. One of the most important is a diagonal line filled with small diamond figures, known as *mlinjon*. In the old days people were convinced that a handwritten batik with *mlinjon* lines on it possessed strong magical powers. This is understandable if one bears in mind the fasting undertaken by the bakit maker in order to concentrate her mind on her drawing, to produce neat and beautiful *mlinjon* lines.

Many patterns contain wings in several forms and sizes, especially in the form of a garuda bird. In wearing winged patterns the wearer is supposed to know that the cloth should be worn with the wings upside down, since those wings refer to God, and sometimes to the king and those who are higher than the wearer.

Traditional Batik Patterns

To illustrate my arguments, I will use the following examples of batik patterns in the Solo (Surakarta) style.

Mugirahayu: this term means 'be blessed by God'. The cloth was hand-drawn in 1920, and it contains a prayer to God, to protect the wearer from evil and bad luck. To give more strength to the prayer, the batik maker has added *mlinjon* lines, though it is not usual. The original pattern has no *mlinjon* lines. This pattern illustrates people's belief that the pattern contains a message and that the *mlinjon* lines possess magical power.

Satria manah: this term means 'a knight achieving his target'. This pattern is teaching us that if an honest man wants to achieve some target, he has to follow the right way. He has not to use bypasses and cunning tricks.

Semen Rama: Rama is the hero in the *Ramayana* epic. This pattern is embedded the well-known lecture of Yosodipuro II, one of the most important Javanese poets around the year 1740. According to him, to be a good ruler or leader, one has to possess eight qualities which are symbolized through eight major ornaments in this pattern. In the *Ramayana* epic, this lesson has been taught by Rama to Gunawan Wibisono, the new kind of *Alengka*. Those eight qualities are abilities to: provide brightness and spirituality; protection; set a good example; maintain respect and order; be a good haven of refuge; punish impartially; be strong and steadfast in making decisions; provide welfare for others.

Bondhet: the major ornament in this pattern is a strange-looking creature, half animal, half bird, with one leg tied and one leg free. This figure symbolizes the ideal relation between husband and wife in a happy marriage. Partly they form a unity and partly they keep their own identities.

Kopi pecah: this name means 'broken coffee beans'. The lesson in this pattern is that sometimes one has to be willing to sacrifice oneself to join a mass, to serve the public interest. We can see it with the coffee bean. In this pattern, the public interest is symbolized by the Garuda figure.

Kawung: this is the name of a palm tree. The figures in this pattern remind us of the fruit of the *kawung* palm, a tree of which all parts are useful. So this pattern symbolizes the hope that the wearer will be as useful to others as the *kawung* tree (Hardjonagoro, 1989). Another interpretation is that this pattern symbolizes fertility, because the figures in the pattern resemble a fruit with seeds in it.

Slobok: this name means 'loose' or 'too wide'. The small crowns in the pattern refer to someone's position in the society. This pattern contains the hope that the wearer of this batik will find a smooth way in his social life and career.

Illustration 4.1   *Mugirahayu*

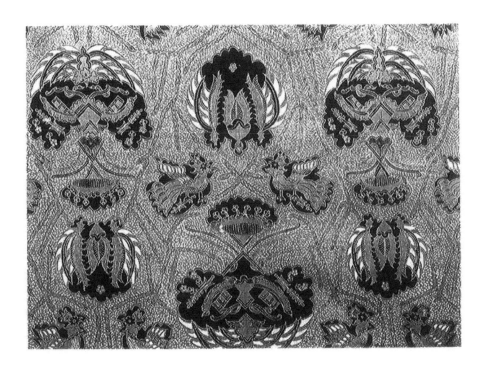

Illustration 4.2 *Satria manah*

Illustration 4.3   *Semen Rama*

Illustration 4.4 *Bondhet*

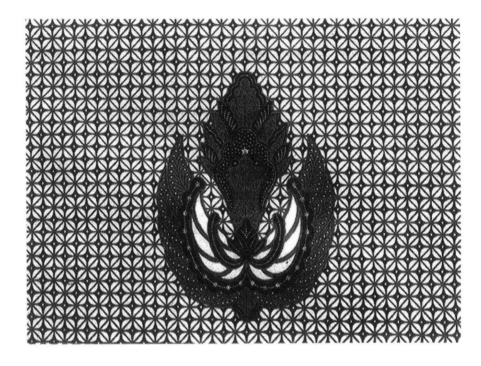

Illustration 4.5   *Kopi pecah*

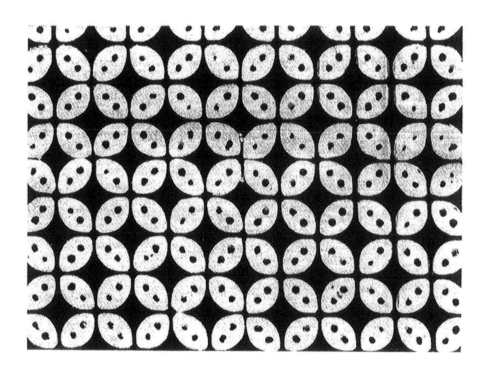

Illustration 4.6   *Kawung*

Illustration 4.7   *Slobok*

As I have mentioned before, batik patterns, even the traditional ones, have undergone modifications, and these new patterns too can have a spiritual content, so that nowadays there are many modified and new batik patterns which also symbolize messages. Below are some examples.

Modified *slobok*: this pattern is designed by Hardjonagoro. The figures in this same *slobok* pattern are rearranged to form diagonal lines, because lines and other geometrical figures add more strength to the spiritual power of the cloth.

Another modified *slobok*: this pattern is also designed by Hardjonagoro and is entirely the same as the former pattern, except that in this *slobok* the background is white. According to the old Javanese etiquette in wearing batik as a cloth, patterns with a dark background are to be used during the day and patterns with a light background in the evening.

Tjuwiri mentool: these tiny point-shaped flowers with stems look very much like the hair accessory or *mentool* that the Javanese bride wears in her coil. Since those *mentools* symbolize fertility, this pattern also contains, among other things, the hope for fertility. This pattern is traditional, but Hardjonagoro has modified the colour, since the original colours were as usual blue, brown and white.

Kodrat: this name means 'nature's way'. The main ornaments in this newly-designed pattern symbolizes man's destiny, to form a pair in life. It is designed by Aniek B. Effendy (1990).

Conclusion

The explanations I have just given are the results of my deliberations with friends. It shows that the spiritual part of batik contains cultural messages. So the knowledge of it is badly needed. Unfortunately, the chain of tradition has been broken since the present generation is unable to hand the knowledge over to those who will come after them.

Based on my observations in Solo and Yogyakarta, there are no batik products which were ever labelled with the name of their patterns. What I and my friends have been struggling for all these years is to persuade people who take part in the batik industry and trade to label the name of the patterns on their batik products. By knowing the name of the pattern, it will be easier to know the message of the batik pattern one has chosen. Hoping that by knowing one's own cultural values one will love it, since in this era of globalization, cultural identity is very important.

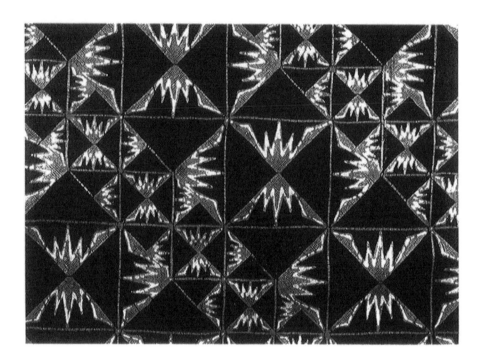

Illustration 4.8a   Modified *slobok*

Illustration 4.8b   Modified *slobok*

Illustration 4.9   *Tjuwiri mentool*

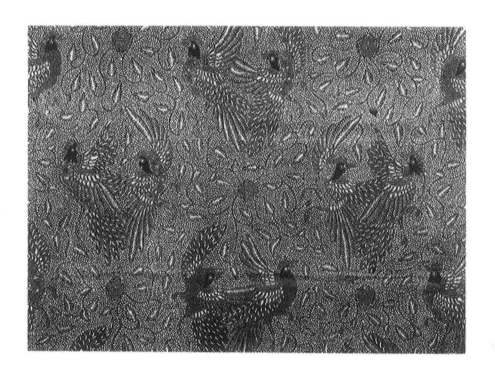

Illustration 4.10   *Kodrat*

References

Djoemena, N.S. (1986), *Batik, Its Mystery and Meaning*, 2nd edn 1990, Jakarta, Djambatan.

Djoemena, N.S. (1990), *Batik and Its Kind*, Jakarta, Djambatan.

Hamzuri (1989), *Batik Klasik* (cetakan ketiga), Jakarta, Djambatan.

Hardonagoro, G.T.S. (1985–97), personal communications.

Naskah (1985), 'Simbolisme dalam Corak dan Warna Batik', *Majalah Femina*, 28, pp. 2–16.

Siswomiharjo, O.K.W. (1992), 'Beberapa Upaya yang Dapat Dilakukan untuk Mengembangkan dan Melestarikan Batik', seminar *Kehidupan Batik Tradisional Indonesia*, Surakarta, Indonesia.

Susanto, S. (1980), *Seni Kerajinan Batik Indonesia*, Balai Penelitian Batik dan Kerajinan – Departemen Peridinustrian R.I.

Van Roojen, P. (1993), *Batik Design*, Amsterdam, The Pepin Press.

Veldhuisen, H.C. (1993), *Batik Belanda 1840–1940. Dutch Influence in Batik from Java, History and Stories*, Jakarta, Gaya Favorit Press.

Veldhuisen-Djajasoebrata, A. (1984), *Bloemen van het Heelal*, Amsterdam, A.W. Sijthoff.

5   Function and Meaning of *Batik-lurik* – A Reconstruction

RENSKE HERINGA

Introduction

As a subject of scholarly endeavour, batik has attracted interest since the beginning of the twentieth century. Although the art is often conceived of as a primarily Central Javanese cultural expression, recent studies have begun to highlight the many different guises of batik in other areas of Java and the archipelago. A wide variety of styles and uses are being mapped.

Batik made along the Pasisir, the north coast of Java, has been given particular scholarly attention (McCabe Elliott, 1984; Veldhuisen, 1993). The variety of Pasisir styles and their historical development since the mid-nineteenth century have been considered in a recent analysis (Heringa and Veldhuisen, 1996). The new data strongly suggest that, in the early nineteenth century, batik from the north coast and from Central Java had much more in common than is generally acknowledged today. Therefore an increasing need is felt to reconsider the development of batik prior to the nineteenth century.

As far as actual examples are concerned, this will be an almost impossible task, because, as far as I know, batik cloth predating the early nineteenth century has not survived. However, Javanese manuscripts and literary sources may contain information for the reconstruction of once-existing styles and function of batik. Another source is the traditional textiles made in villages in the hinterland of Tuban, the ancient port town on the northeast coast. Although the hand woven cloths have been made and worn for centuries by village women inside the area, they were not noticed by a wider public until the mid 1970s. This was due to the initiative of the Department of Industries to revive the craft. This undertaking has been quite successful, although it must be noted that, inadvertently, the appearance of the cloths that since have been made for the outside market has undergone enormous changes. Only a small amount of textiles made for use by the villagers themselves still reflect the traditional standards described in this paper. As the only survival of a batik tradition predating the commercialization of the production process on the North coast, the textiles from Tuban offer insights in the erstwhile function of village textiles – both batik and woven cloths – as identity markers.

Besides weaving *lawah*, the base cloth for batik, the village women in Tuban continue to master a variety of other techniques. A comparable situation was found throughout Java during the early nineteenth century by Thomas Stanford Raffles: 'The men of all ranks pride themselves on the beauty of a cloth woven by their wife, mistress or daughter' (Raffles, 1817, p. 86). However, the variety of textiles available in Tuban offers a much more complex picture than that presented by Raffles.

For a proper understanding of the social function of the textiles, it must be noted that in earlier days each particular technique was associated with the dress of men of specific status groups within the community. Wives indicated their relation to their husband's group by using the appropriate technique and also wearing it themselves. Today the men have replaced the sarongs made by their wives with trousers and jackets made out of industrial fabrics, but many women continue to adhere to the old social distinctions. Batik is only allowed to be made and worn by the village elite, the landowners out of whose ranks the village leadership is chosen. The rest of the population is restricted to a variety of striped and checked textiles generally known as *lurik*. A similar division used to be in force in Central Java, where batik was reserved for the aristocracy, while *lurik* primarily served as dress for the common people.

In this chapter, updated research on the visually least spectacular among the Tuban textiles, *batik-lurik*, will serve to point out connections between traditions of the North Coast and Central Java. First, information from archival sources and a museum collection corroborates the extensive use of *batik-lurik* in other regions on Java until the last quarter of the nineteenth century. Then, a close look at the technique, the motifs and colours of *batik-lurik*, shall lead to an analysis of its present-day function. A comparison with other aspects of Javanese culture, including Central Javanese batik motifs, indicates a role for *batik-lurik* which in earlier times communicated identity far beyond the villages.

Historical Sources

The earliest sources mentioning *batik-lurik* date to 1845. Sample books, which used to belong to a nineteenth century Dutch textile printing factory, show a wide range of samples of real *batik-lurik* and of machine printed imitations in many shades of reddish-brown, blue and black. The accompanying correspondence shows that huge quantities of *batik-lurik* copies were produced by the Dutch textile industry from the 1840s and were exported to Java. The extent of the demand for *batik-lurik* all over Java is thereby abundantly established. The shoddy textiles were sold cheaply and probably contributed to the general disappearance of handmade *batik-lurik*. Whatever the case, after 1860 *batik-lurik* gradually disappears from the sample books. Around 1880 the market for the imitations had disappeared.

The samples consist of plain machine-woven cotton of the cheapest quality covered with small floral printed motifs. In many cases they are embellished with the typical *kepala tumpal* and floral borders in line-drawn batik technique. The only known example of real *batik-lurik* is a textile from Lamongan in the collection of the Ethnography Museum in Leiden. It shows exactly the same lay-out as that encountered in the sample books, the type of *kain panjang kepala tumpal* which used to be worn all along the north coast. Evidence for the use of *batik-lurik* in other areas on Java is further strengthened by a description of dress forms in the Kediri area (southern part of East Java) by the Reverend Poensen. In 1877 he refers to hip wrappers worn by village women for special occasions as: 'lurik which has been batiked in the checks, and which is called batik-lurik.' Conflicting information comes to us from a much later period; in approximately 1930 Ostmeier mentions the local dress-style of several villages near Demak (central area of north coast Java) as 'red cloths with small white starlike flowers' (Ostmeier, nd, p. 262). This rather meagre description may refer to several techniques: either *batik-lurik*, which was not recognized by the author or woven cloths with small woven-in flowers or even to an ordinary batik cloth. Nevertheless, it can be concluded that *batik-lurik* was worn and made in regions other than Tuban.

Technique

The base cloth consists of several types of locally hand spun, checked cotton material. Although the cloth technically belongs to the *lurik* type, the villagers call it *klontongan*, empty shell, explaining that the 'contents', the meaningful decorations which give the cloth-and its owner-their identity, have not yet been applied. Several types of checks in black or red on white each serve as a grid for the time-consuming waxing of particular motifs. Most of the small geometrical motifs are made up entirely of series of dots.

After immersion in a cold dye bath and removal of the wax in boiling

Illustration 5.1   Waxing *batik-lurik*

Illustration 5.2   **Dutch *batik-lurik* imitations**

water, the motifs stand out against the darker background. Those on the blue ground show small flowers in white and red, as the wax has been applied to both the white areas and the red checks. The white dotted motifs are enclosed by the black checks shimmering through the red base. The designs cover the cloth in a dense, evenly spread layer, which changes into a series of separate motif bands crossed by warp-stripes in the *kepala* of a *sarung* or the section of a *kain panjang*. Most probably the two types of hip wrapper were the only formats in which *batik-lurik* was made.

Originally, the technically simple, geometrical motifs may have been applied with a pointed stick or a piece of bamboo, instead of the *canting*, the spouted waxing pen which is used today. Some of the motifs resemble those of *kain simbut*, the rice paste cloths of West Java, which were also drawn with a flattened piece of bamboo. A series of geometrical motifs denoted by the same names is known in the Central Javanese idiom. Here the dotted areas have been modified into line drawings, that can only be drawn with a *canting*. It appears logical that the *batik-lurik* version of the motifs are of an earlier origin.

Colours

The first important aspect of *batik-lurik* are its colours: either blue-black and red or the contrast between dark and bright. The village women explain that dark checks need a dark cover. This is related to the concept of cosmic totality in which brightness of day and darkness of night form an indivisible unity as they follow one another. During the day, the sun climbs to its highest point in the south and finally sets in the west. During the night it returns through the north to the east. This concept has developed into a relation between the colours of the cloths and the position of a person in time or space. First the use of bright or dark cloths, red and dark blue, indicates the different phases in a woman's life cycle. After the end of life the body, covered by a black *batik-lurik,* is laid out in the centre of the house. Together the three colours represent the generations of the living and the ancestors who are thought to dwell in the same family compound.

A second sign of identity is contained in the relative strength of each colour, indicating the village to which each woman belongs. Early in the morning each woman on her way to the market can be recognized at first sight. Those from the north are clad in blue-black. Women from the east wear *abang hati*, bright red. Those from the west show their own version of base checks and motifs: *grompol paccar* for the motif from the east, *grompol songo* for the clusters of 16 dots in the motif from the west. Colour and motifs have kept their importance as markers of identity for the village women.

Meaning and Function of the Motifs

Less is known about the meaning of the motifs and their function. Initially the village women consistently named the few *batik-lurik* patterns, but were unable to offer any explanations as to their meaning or use. The translation of the Central Javanese motif names given by Jasper in his classic book *De Batikkunst* does not clarify the matter (Jasper and Pirngadie, 1916, pp. 39, 164). However, once the villagers considered the geometrical forms as abstractions of ideas, some of the motifs could be explained as pictograms of spatial concepts well-known throughout Java. Associations with other cultural aspects led to further helpful suggestions.

The *grompol* motif, consisting of clusters aligned in alternating groups of five and nine dots, was said to resemble a diagram of groups of four and eight villages, one in each cardinal direction, with a fifth or a ninth in the centre. Although the villagers do not have any particular term for these concepts, it is known in the literature as *mancapat* and *mancalima*, and refers to the ideal model according to which villages in many areas of Java used to form a cooperative system.

The *cuken* motif is also found painted on the entrance gates to the villages, and is explained by the villagers as an ancient symbol of the yearly agricultural cycle. In Jasper's book it is called a solar wheel. In the *kijing miring* motif we see rows of triangles aligned diagonally across the cloth. *Kijing miring* means (fallen) gravestones, which relates the motif to ascending lines of ancestors and to the ancestral graves. At this site of vital ritual importance the yearly ceremonial food exchanges, the *manganan*, initiate the beginning of the new agricultural cycle.

Although gradually a partial explanation emerged as to the meaning of the previous motifs, the villagers still do not remember by whom or for what occasion they may have been worn. One motif name, *kasatriyan*, offers an answer to this question. It means 'for warriors', a function which may at first sight seem curious in a village setting. Nevertheless, some of the villagers relate how, in the olden days, their ancestors were given the special task of joining the ruler of Tuban in his campaigns. In those long-gone days, the motif possibly served as the warriors' special marker of identity and badge of

Illustration 5.3   *Kijing miring* (Central Java)

Illustration 5.4   *Cuken* **motif (Central Java)**

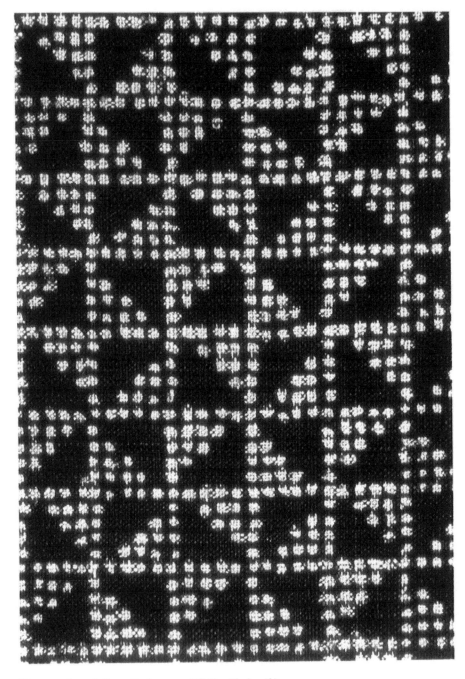

Illustration 5.5   *Cuken* motif (*batik-lurik*)

honour. Once the Tuban ruler lost his political independence in the eighteenth century, and the campaigns came to an end, the special function of the motif disappeared. The motif remained but its meaning was forgotten. Interestingly, a woven version, like the *batik-lurik kasatriyan*, consisting of alternating bright and dark checks, also forms part of the Tuban textile vocabulary. The question remains whether the function of the two textiles may have been different.

The last *batik-lurik* motif to be mentioned here is *kawung*, which is made in an intermediate form somewhat closer to the Central Javanese line patterns. Although it is enclosed within rather larger checks, it consists of lines instead of dots.

A closer look at the function of all of these line-drawn 'cousins' to the *batik-lurik* patterns, and a comparison to the hip wrappers worn by *wayang* (shadow theatre) figures shows that they are worn by various *wayang* puppets. Although geometrical motifs are also worn by other *wayang* figures, Semar and his sons, the Panakawan, servants and followers of the Pandawa, the heroes in the *wayang* stories, are generally shown wearing patterns that are also found in *batik-lurik*. Older puppets show the geometrical motifs, in this case *kasatriyan* for Semar, the father and highest in rank, and *kijing miring* on Gareng and Petruk, the sons. Bagong, in two different moods, may be shown wearing *cuken* and *kijing miring*. This suggests there was indeed a distinction or even a ranking system between the different motifs. Such a distinction has disappeared in the case of more recent Panakawan, from Yogyakarta, who all wear *kawung*. The Panakawan are considered to be the ancestors of the common Javanese, the *wong cilik*. Thus the use of the *kasatriyan* motif by the followers of the ruler of Tuban seems quite appropriate. The question remains whether the different patterns each indicated a particular task or rank. The overlay of high-status batik on a commoner's *lurik* cloth seems to agree with the position of commoners who have been elected to a special task by the ruler. Due to eighteenth century social-political changes, *batik-lurik*'s erstwhile role as status marker in a wider sphere has lost its relevance for the villagers.

References

Elliott, I. McCabe (1984), *Batik: Fabled Cloth of Java*, New York, Clarkson N. Potter.

Heringa, R. and Veldhuisen, H. (1996), *Fabrics of Enchantment*, Los Angeles, Los Angeles County Museum of Art.

Jasper, J.E. and Pirngadie, M. (1916), *De Inlandsche Kunstnijverheid in Nederlandsh-Indië, Vol III: De Batikkunst*, S'Gravenhage, Mouton.

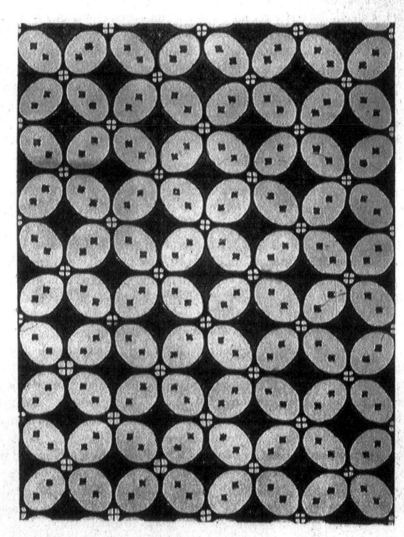

.2. Ragam hias: kawung seru latar putih. Djawa Tengah

Illustration 5.6   *Kawung* **motif (Central Java)**

Raffles, T.S.B. (1817), *The History of Java*, London, Black, Parbury and Allen.

Veldhuisen, H.C. (1993), *Batik Belanda 1840–1940: Dutch Influence in Batik from Java, History and Stories*, Jakarta, Gaya Favorit Press.

6 The Philosophy and Meaning of Classic Batik Patterns of Central Java

S. HERTINI ADIWOSO

For better understanding of the significance of batik patterns, it is necessary to locate them in their historical and cultural context. Let us take an example. One ought to be careful with one's words. Spoken words are able to express a hope, a prayer but also a curse. It was not right to say *hati-hati* (take care) to persons who were leaving since saying those words is like anticipating something bad or misfortune. One should say *selamat* (good luck).

Not only with words one has to be careful, but also with behaviour. Never hurt somebody's feeling, since it can also happen to you. Nothing in this world is eternal, it always changes: what you do to others can be done to you.

To explain the meaning of all the classic patterns is impossible. So I chose for this paper the *sidos* most known and worn in connection with the life cycle.

The main and important batik centres are Surakarta (Solo) and Yogyakarta (Yogya). The batiks of Solo dominate the coastal areas of north Central Java, while the southern areas and Banyuman are very much influenced by the Yogya batiks. Solo batik is brownish yellow and the nuances melt smoothly into each other. The filling of open spaces, in this case the *ukel*, are fine. The character of the Solonese can be read in it: soft, fine, elegant and almost feminine.

While the colours in the Yogya batik are clear and strong, the outlines of the patterns are very sharp. The contrast between white and brownish black is very clear: strong and courageous in character.

Sido is derived from *sidoho* (an exclamation) meaning to hope or to wish. *Sido asih* means 'may the wearer of this cloth be loved'; *sido mukti* – 'may they be bestowed with worldly comfort'; *sido mulio* – 'may the become honoured and highly respected'; and *sido luhur* – 'may they be noble of mind and conduct, and highly respected because of their wisdom'.

64

With the exception of the Yogya *sido mukti*, the clothes are divided into squares, not by straight lines, but by graceful wavy lines. Since life is never straight, there are always up and downs and twists and turns. In each square is a figure, a house, a wing, a butterfly, and a flower.

Everyone needs a house, a shelter for life. A roof is depicted clearly, a shelter against the weather, sun and rain, covering the *pendopo* (veranda). The house may resemble a small Hindu temple the way the artist sees it.

The wing expresses 'may the wearer fly and reach his destiny'. The butterfly symbolizes continuation of the family line. The butterfly only lives to lay eggs, after which it dies. It also a symbol of fertility. Flowers express 'may life be filled with flowers and fragrance'. Sometimes another figure is present, a cross, or the points of a compass, the direction of the winds. Whatever wind comes, one has to stand firm in one's place, having one's own opinion. All those figures might be seen in the *sidos* of Surakarta and Yogyakarta, excepting the Yogyakarta *sido mukti* mentioned above.

The spaces in the squares with the main figure in the centre are filled with small figure *isen-isen* (filling) named *ukel*. *Ukel* means knot, a woman's hairdo; the traditional way is also *ukel*, a kind of intertwining. The Surakarta *ukel* is fine and small, whereas the Yogyakarta variety is bigger.

In *sido asih*, the squares are filled alternately with *ukels*, black or light-coloured. In old days this cloth would be worn by the young men and women at their betrothal, although they would not meet each other: everything was arranged by the parents and family. Later when customs became less strict, the *sido adih* was worn by the young man and girlfriend at the official announcement of their engagement to both families.

The *ukel* in the squares symbolized the intertwining of the families, as well as the small problems to be faced and resolved. The black and light-coloured squares comprise the black of the Yogya batik and the light of Solo with flowers in the corners. The white means the clear mind of the parent and the black their calmness to guide the young people to prepare for married life.

The *sido mukti* expresses a hope to be well cared for and bestowed with worldly comfort, and is worn by bride and groom on their wedding. This is the Surakarta *sido mukti*. Usually the parents of the bride and of the groom wear the *truntum*, to express the hope of becoming grandparents.

Sido mulio and *sido luhur* are similar but not the same. *Mulio* is honoured and is addresses as 'your honour'. *Sido mukti* is in general light-coloured. The Surakarta *sido mulio* is yellowish brown, while the Yogyakarta *sido mulio* is dominated by white. *Ukel* is not found any more. The house, wing, flowers and sheaf of rice are shown on the Yogya *sido mulio*. Some of this batik cloth has only wings and butterflies. Flowers decorate the corners like a garland, symbolizing that life has reached a stage of fragrance all around.

Sido luhur is black, the colour of dignity and the calmness of the mind. It is divided into squares with corners bearing flowers or curls (but not in *ukel* form), a small bird, the phoenix, wings, a house, a bee (usefulness), flowers, and the lotus. Some batiks, especially the Yogya *sido luhur*, have a complete wing figure, *sawat*. *Sido luhur* is usually worn by older people who have experienced everything in life. They have overcome the problems and turbulence of life and have become wise through age and experience, because wisdom comes with age.

Yogyakarta has always been different; a monarchy in a republic. Their *sido mukti* is as distinctive as Yogya itself. The cloth is not divided into squares, but is covered with *ukel*. It indicates that although problems are everywhere, the couple will hopefully find happiness and a worldly comfort. The wings – the four directions of the wind – stand for what has past. The flowers are everywhere to fill their life. The pheasant is the continuation of the family line. Coconut leaves represent the coconut tree, a symbol of usefulness. The *sawat*, two wings and tail, give shelter.

The batik cloths symbolize hope, prayers, status, dignity and wisdom. No wonder Iwan Tirta called it the magic cloth.

Although some *sidos* are for special occasions, they may also worn on other occasions, as long as one is careful. Wearing a batik cloth is not as easy as it seems. To wear batik cloth is to take care; the figures must not be upside down, except when the figures are in two directions. The line should be in the right direction – up or down. Also one has to to know exactly what to put on at each occasion.

Our batik scholar and expert has expressed it succinctly: when, where and who? I would like to add a mystical consideration. For practical reasons the ladies wear this cloth as a skirt, although not tailored to the body or fitted with a zipper. The artist is the tailor. The old traditional and right way (for me) is to wear batik cloth by wrapping it round the lower body neatly and tightly, showing the shape of the body, but draped so that one can walk easily and gracefully. Tying the magic cloth in this way is an art itself.

7   Textiles in Ancient Bali

I WAYAN ARDIKA

Introduction

In the late ninth century, inscriptions in the Old Balinese language began to appear in Bali. Most of these inscriptions recorded the establishment of *sima* or the transfer of taxes and labour rights by a ruler or highly placed taxing authority to the religious foundations. Crafts specialists were among taxpayers mentioned in these inscriptions.

The aims of this chapter are to describe the history of textiles in ancient Bali as well as the status of crafts specialists in Balinese society between the ninth and eleventh centuries. This period was crucial in the economic history of Southeast Asia and was known as a regional 'trade boom' (Christie, 1996). The increase of trade activities in the South China Sea and Indian Ocean during the period might also have stimulated the appearance of craftsmanship in Balinese society. In the commercial development model, increases in specialization and exchange are seen as integral parts of the economic growth. The economic growth encourages individuals to avail themselves of the efficiencies of specialization and exchange, and as the division of labour becomes more elaborate, so social complexity increases (Brumfiel and Earle, 1987).

In the political model, local rulers have an important role in organizing specialization and exchange. It is proposed that political elites consciously and strategically employed specialization and exchange to create and maintain social inequality, to strengthen political coalitions and to fund new institutions of control, often in the face of substantial opposition from those whose well-being is reduced by such actions (Brumfiel and Early, 1987, p. 3; Earle, 1987; Earle, 1994; Peregrine, 1991). There are several versions of how the rulers controlled certain products through the sponsorship of craft production or trade. In the first version, monopoly over foreign commerce is regarded primarily as a source of profit for the ruler, a source of income that can be invested in an array of mechanisms for augmenting the leader's power. In the second version, a ruler achieves coercive power over a population by monopolizing certain food crops, tools or weaponry. The third version of the political model suggests that the control and manipulation of wealth is a key factor to building political power. Wealth can come into play in the initial stages of social ranking. Finally, control over prestige goods or wealth, when combined with a regional market system, could provide a means of supporting administrative and craft specialists working for the state.

Crafts specialists and exchange is seen as a part of an economy based upon redistribution. In a region of high resource diversity, where locales are optimally suited to producing different things, specialization and redistributive exchange would confer substantial benefits.

Crafts Specialists

Several crafts specialists were mentioned in 33 Old Balinese inscriptions, dating from the ninth to eleventh centuries including: smiths of bronze, gold, silver and iron, dyers and other textile supporting industries, potters, basket makers, carpenters, ship builders, stone workers, tunnel workers for irrigation purposes, sculptors, bricklayers and artists. These craft specialists might have been part-time or semiprofessional: manufacturing and processing industries operated in some households, and they had to pay taxes.

In Old Javanese society, textile production and other activities related to the processing and use of dyestuffs were referred to by the term *misra* (a Sanskrit term meaning 'various'). The term *misra* appears to have encompassed a number of part-time, semiprofessional manufacturing and processing industries which operated in some farming households in Javanese villages of the time (Christie, 1993, p. 185). *Misra* (part-time specialists) and *samwyawahara* (full-time specialists) categories were not mentioned explicitly in Old Balinese inscriptions dating from the ninth to eleventh centuries. However, the variety of crafts specialists mentioned in Old Balinese inscriptions may be classified into two categories, namely part-time and full-time crafts specialists. The part-time specialists (*misra*) in Old Balinese society consisted of dyers in red and blue, potters, basket makers and lime manufacturers. On the other hand, the full-time specialists included smiths, carpenters, sculptors, musicians and dancers.

Craft specialists concerned with textile production might have special status in Old Balinese society. Clothing is needed for personal ornament by rulers and society in general. Old Javanese inscriptions indicate that varieties of clothes were related to social status and gender of people in the society

(Christie, 1993, p. 194). These phenomena were not explicitly mentioned in Old Balinese inscriptions.

Apart from craft specialists, artists seem to have special status in ancient Bali. At the beginning of the eleventh century there were two types of artists mentioned in Balinese inscriptions, namely those who were attached to the rulers (*i haji*) and the independent ones (*ambaran*). Artists who were attached to the rulers were normally paid more than those who were independent. It is likely that attached artists were better in quality than those who were independent. Artists that were mentioned in Old Balinese inscriptions include: singers, flutists, drummers, gamelan players, mask dancers and puppeteers.

The existence of art and craft specialists in ancient Bali seems to be related to the development of social stratificiation and large political and religious institutions.

Crafts Productions

According to Brumfiel and Earle (1987) craft production can be divided into two categories, namely subsistence goods and wealth. The subsistence goods comprise food and drugs, and the productive technology is used to satisfy basic household needs. Wealth includes primitive valuables which are used in display, ritual and exchange, as well as special, rare and highly desired subsistence products.

Based on these categories, several craft products that were mentioned in Balinese inscriptions dating from the ninth to eleventh centuries can be categorized as subsistence goods such as pottery and baskets. These products would have been needed by every household at that time. However, it should be noted that potters (*mangdyun*) and basket makers (*manganyam*) were not mentioned explicitly in the Old Balinese inscriptions. This might be due to the low status of these craft specialists in the society of that period.

Other products such as metal objects and garments could be categorized as wealth. As far as metal is concerned, its raw material is absent in Bali. Archaeological data indicate that Bali has been involved in long distance trade in metals since prehistoric times (Ardika, 1987 and 1991). Due to the rarity of raw materials of metal in Bali, it can be argued that metal objects must be high-prestige goods in Balinese society. Apart from land and animal husbandry, gold, silver, copper and bronze drums were mentioned as important inheritances in Old Balinese inscriptions.

According to Christie (1993, pp. 184, 194), certain types and colours of clothes during the tenth century had special meanings in Old Javanese society. In other words, the quality and colour of clothes referred to the status of their owners. It is interesting to note that during the late tenth century when the Old Javanese language appeared in Balinese inscriptions several types of clothing were mentioned including *wdihan, basahan, kurug, wastra, dodot, kampuh* and *sinjang* (Stuart-Fox, 1993, pp. 90–1). It is likely that similar phenomena, which occurred in Javanese society, might have also appeared in Bali during the tenth century. Special type and colour of garments or clothes might have also been categorized as wealth in Old Balinese society.

Other craft productions such as statues and religious objects were very important in Old Balinese society and could be categorized as wealth. Sculptors (*anatah*) and stone builders (*undagi batu*) were among the other craft specialists mentioned in Balinese inscriptions.

Textiles and Trade

The appearance of early bronze metallurgy in Bali, which lacked the necessary raw materials, copper and tin, suggests that Bali was involved in inter-island trade of metals (Ardika, 1987). Tin might have been obtained from the Malay Peninsula, and Bangka or Baliton in the western part of Indonesia. Copper occurred in Java and Sumatra.

Apart from metals, the discovery of rouletted-ware and other types of Indian pottery in Sembiran, northeastern Bali, also suggests that contacts between Bali and India might have begun around 2,000 years ago. In terms of forms, decorations and raw materials of the Indian pottery found in Sembiran are very similar to those found in Arikamedu in Tamil Nadu and other Indian sites (Ardika, 1991; Ardika and Bellwood, 1991; Ardika et al., 1993; Ardika et al., 1997). The appearance of Indian pottery in Sembiran might suggest direct contact between Bali and India at the beginning of the first millennium. So far, Sembiran has produced the largest collections of Indian pottery in Southeast Asia.

Indian written sources indicate that gold and spices were the main commodities which attracted the Indian traders to Southeast Asia. Spices such as cloves and sandalwood are native to Moluccas and Nusa Tenggara Timur, in the eastern part of the Indonesian archipelago. These products might have traded through Sembiran to the western part of the archipelago. In other words, Sembiran could be an ancient port located on the spice trade routes linking eastern and western Indonesia.

Supposing that gold and spices were the main commodities sought by the Indian traders, the question also arises of what were the commodities brought in by the Indian traders to Southeast Asia, especially Bali. Many scholars believe that textiles were the main products which were traded from India to Southeast Asia and Africa (Guy, 1996; Barnes, 1996). Clothing materials such as cotton and dyes as well as techniques of textile manufacturers could derive from India and could have been introduced to the people of Southeast Asia in the first millennium BC, prior to which time bark-cloth and textiles of plant fibres other than cotton were presumably the primary clothing materials. However, recent research has indicated that *Gossypium hirsutum*, a New World species, may have existed in insular Southeast Asia before the introduction of Indian cottons (Stuart-Fox, 1993, pp. 86–7). Apart from Indian pottery and beads, textiles and cottons, as well as dyes, might have also reached several ancient ports or coastal sites in Bali such as Gilimanuk, Julah (Sembiran) and Banua Bharu.

Cotton (*kapas*) and cotton yarn (*benang*) were first mentioned in the inscription of Bebetin (Banua Bharu), which have been dated at AD 896 (Goris, 1954, pp. 54–5). The inscription of Sembiran B (AD 951) mentions a cotton plantation (*parkapasan*) in the Julah area on the northeastern coast of Bali (ibid., p. 72). According to the inscription of Sukawana D (AD 1300), an important cotton-growing area was the land to the east of Sukawana, between Panusuran and Balingkang. Kintamani people bought this cotton and marketed it in the north coast villages (Stuart-Fox, 1993, p. 87).

An interesting piece of information regarding cotton production was mentioned in the inscription of Kintamani E (AD 1200). The inscription records the right of the people of Cintamani (Kintamani), as in time past, to trade cotton to the villagers of Les, Paminggir (modern Tejakula), Hiliran (unidentified), Buhundalem (modern Bondalem), Julah, Purwasiddhi (unidentified), Indrapura (modern Depaha), Bulihan and Manasa (near Sinabun?). All villages were on the northern and northeastern coast of Bali where presumably cotton was spun and woven (Stuart-Fox, 1993, p. 87). The villagers were also allowed to trade safflower (*kusumba*), amongst other things. It is also mentioned that the people of the villages of *Wingkang Ranu* (the villages around Lake Batur, namely Abang, Batur, Bwahan, Kedisan and Trunyan) were forbidden to sell cotton (*kapas*) to the villages on the northern coast of Bali.

Old Balinese inscriptions mention two types of dyeing activities, namely *mangnila* (indigo dyeing) and *mamangkudu* (morinda dyeing). Several Balinese inscriptions also mention taxes or levies on these activities. *Mangnila* derived from the Sanskrit word *nila*, namely indigo dyeing. According to Stuart-Fox (1993, p. 88) the term *mangnila* might suggest some connection between Bali and India, particularly in the trade of indigo dyeing and the influence of the indigo dyeing techniques. Gujarat in west India was known from the earliest times as the centre of indigo cultivation, which exported indigo to Persia in the sixth century and to Sung Dynasty China. Ethnographic data indicate that indigo was still grown as a secondary crop on rice land in north Bali until at least the late nineteenth century. Various ritual restrictions are attached to indigo production and dyeing.

Old Balinese inscriptions indicate a well-organized and well-regulated trade network linking north coast villages and mountain villages in the Kintamini area. Cotton was an important commodity in this system. Trade in cotton seems to have been regulated and controlled by the rulers: special taxes were also imposed on cotton and cotton yarns. Taxes on dyestuffs also indicate that the ruler gave much attention to textile crafts.

Sources form the sixteenth century show that Balinese cloth was exported particularly to the Moluccas. The trade in simple cotton cloth continued until the nineteenth century.

It is worth noting that along the north coast, cotton, indigo and safflower all remained important crops until the end of the nineteenth century, suggesting a continuity of 1,000 years or more (Stuart-Fox, 1993, p. 92).

Textiles and Status

As already noted, craft production is needed by all members of the society, whether elite or commoner. The status of crafts specialists may depend on their products. Craft specialists who produced subsistence goods might have had lower social status than those who produced wealth.

Based on this hypothesis, local rulers or elite groups have important interests in organizing and controlling craft specialists. By doing this, the local rulers would control certain products through the sponsorship of crafts production or trade in order to increase their power. The political model suggests that the control and manipulation of wealth is a key factor in building political power. Wealth can come into play in the initial stages of social ranking. Control over prestige goods or wealth, when combined with a regional market system, could provide a means of supporting administrative and crafts specialists working for the state.

Balinese inscriptions dating from the ninth to the eleventh centuries mention several high functionaries in connection with craft specialists such as *Samgat, Nayaka, Juru* and *Tuha*. The *Samgat Juru Mangjahit Kajang* and *Samgat Mangjahit Kajang* were mentioned in the inscriptions of Manik Liu AI, C and Manik Lui BI (Goris, 1954, pp. 74–5; Ardika and Sutjiati, 1996, pp. 34–6). These functionaries may have related to certain types of clothes.

Several high functionaries on textiles mentioned above might have related to fine clothes (*kajang* and *cadar*). In the inscription of Pengotan AI dated at AD 924, a *Tuhan Cadar* was mentioned. The *Tuhan Cadar* was probably an officer concerned with fine clothes.

These pieces of information also indicate control over textiles by Balinese rulers in ancient times. As in Old Javanese society, it is likely that special cloths might have related to the high status of the owners in ancient Balinese society. However, it was not mentioned explicitly in Balinese inscriptions between the ninth and eleventh centuries.

Conclusion

Trade in cotton and dyestuffs, as well as textile industries in ancient Bali, might have been controlled and regulated by the rulers. Textiles can be categorized as wealth which indicates the status of the owners.

Craft specialists on textiles as well as others were controlled and taxed by their rulers. They seem to have been part-time professionals. The status of crafts specialists in society is related to their products. Craft specialists who produce wealth (e.g. textiles and metal objects) would have higher status than those who manufacture subsistence goods. This is due to the importance of craft production, which was needed by the elite or rulers in ancient Bali.

References

Ardika, I.W. (1987), 'Bronze artifacts and the rise of complex society in Bali', thesis, Canberra, Australian National University.

Ardika, I.W. (1991), 'Archaeological Research in Northeastern Bali, Indonesia', dissertation, Canberra, Australian National University.

Ardika, I.W. and Bellwood, P. (1991), 'Sembiran: the beginnings of Indian contact with Bali', *Antiquity*, 65, pp. 221–32.

Ardika, I.W. and Sutjiati Beratha, N.L. (1996), *Perajin pada masa Bali Kuna Abad IX–XI Masehi*, Laporan Penelitian, Denpasar, Fakultas Sastra Universitas Udayana.

Ardika, I.W., Bellwood, P., Eggleton, R.A. and Ellis, D.J. (1993), 'A Single Source for South Asian Export-Quality Rouletted Ware', *Man and Environment*, Vol. XVIII, No. 1, pp. 101–9.

Ardika, I.W., Bellwood, P., Sutaba, I.M. and Citha Yuliati, K. (1997), 'Sembiran and the first of Indian contacts with Bail: an update', *Antiquity*, 71, No. 271, pp. 193–5.

Barnes, R. (1996), 'From India to Egypt: The Newberry Collection and the Indian Ocean Textile Trade', paper presented at the conference *Seafaring Communities in the Indian Ocean 4th century BC – 15th century AD*, 30 June–5 July, Lyon.

Barret Jones, A.M. (1984), *Early Tenth Century Java from the Inscriptions*, Dordrecht, Holland, Fris Publications.

Brandes, J.L.A. (with N.J. Krom) (1913), *Oud Javaansche Oorkonden*, VBG No. 60.

Brumfiel, E.M. and Earle, T.K. (eds) (1987), *Specialization, Exchange and Complex Societies*, Cambridge, Cambridge University Press.

Christie, J.W. (1993), 'Texts and Textile in Medieval Java', *BEFEO*, 80 (1), pp. 181–211.

Christie, J.W. (1996), 'The Banigrama in the Indian Ocean and the Java Sea during the early Asian trade boom', paper presented at the conference *Seafaring Communities in the Indian Ocean 4th century BC – 15th century AD*, 30 June–5 July, Lyon.

Earle, T.K. (1987), 'Specialization and the production of wealth; Hawaiian chiefdoms and the Inka empire' in E.M. Brumfiel and T.K. Earle (eds) (1987), *Specialization, Exchange and Complex Societies*, Cambridge, Cambridge University Press.

Earle, T.K. (1994), 'Wealth Finance in the Inka Empire: Evidence from the Calchaqui Valley, Argentina', *American Antiquity*, 59 (3), pp. 443–60.

Goris, R. (1954), *Prasati Bali I & II*, Bandung, Masa Baru.

Goris, R. (1960), 'The Position of the Blacksmith' in W.F. Wertheim (ed.), *Bali Studies in Life Though and Ritual*, The Hague and Bandung, W. van Hoeve Ltd.

Guy, J. (1996), 'Rama, Rajas, and Courtesans: Indian Figurative Textiles in Indonesia', paper presented at the conference *Seafaring Communities in the Indian Ocean 4th century BC – 15th century AD*, 30 June–5 July, Lyon.

Peregrine, P. (1991), 'Some political aspects of crafts specializations', *World Archaeology*, Vol. 23 (1), pp. 1–11.

Stein Callenfels, P.V. van (1926), *Epigraphia Balica*, I VBG, LVI, pp. III–70.

Stuart-Fox, D.J. (1993), 'Textiles in ancient Bali' in M.-L. Nabholz-Kartaschoff, R. Barnes and D.J. Stuart-Fox (eds), *Weaving Patterns of Life, Indonesian Textiles Symposium 1991*, Basle, Switzerland, Museum of Ethnography.

8 The Classical Batiks of Jambi

FIONA G. KERLOGUE

It was Sir Stamford Raffles who first introduced the art of batik to the English-speaking world, through his account of the process in *The History of Java* (1817). Since then, interest in Indonesian batik has focused largely on the batiks produced in the courts of central Java. The batiks of Jambi in Sumatra are much less well-known, and although Jambi produced, and still produces, fine and distinctive cloths with their own unmistakable character, many writers have doubted that Jambi ever produced batik.[1] This ignorance of the existence of batik in Jambi reflects Jambi's historical position as something of a backwater in Indonesia, and in particular the fact that Jambi resisted Dutch rule until as late as 1904. Jambi's staunch resistance meant that Dutch officials had little opportunity to collect textiles from the region and even less to observe production methods. The result has been a distortion in the representation of Jambi textiles, including batik, songket and a range of other cloths traditionally produced there, both in literature about Indonesian textiles and in museum collections. Misapprehensions have been compounded, since scholars consulting the standard sources for information about Jambi textiles found nothing there and concluded that Jambi did not produce its own textiles. Later, doubtful opinions based on very sketchy information have been repeated and come to be regarded as accurate. This chapter is intended as a start in setting the record straight.

The Adam Cloth

In 1927, a Dutch ethnographic researcher and photographer, Tassilo Adam, presented a Jambi batik cloth to the Ethnographic Department of the Colonial Institute in Amsterdam.[2] He had acquired the cloth in the Pasemah Highlands of South Sumatra in 1921 and had seen similar headcloths being worn in upriver Jambi. Later he had been surprised to come across a workshop in Dusun Tengah, in the district known as Seberang on the north bank of the Batanghari opposite Jambi city, where women's headcloths of exactly the

75

same type were being manufactured. Batik making in the village seemed to be a long-standing and still flourishing practice, carried out by local Malay women, not Javanese. These distinctive headcloths (*selendang*) with their large central lozenge are in design reminiscent of the *kain simbut* of Sunda, but although lozenge shapes do appear in Javanese batik, they do not appear in cloths of this size and shape. A Jambi *selendang* of this type is shown in Illustration 8.1.

The Production Process

In Jambi batik making, the natural dyes used differ from those used in Java, and this points to a long history of batik making in Jambi (Heringa, 1994, p. 26). The process described by Adam includes a number of materials of which there is no mention in accounts of Javanese practice. The finely-chopped wood of the *lembato* tree was used to make an infusion which resulted in a yellow ground colour.[3] There are no reports of the use of *lembato* in Java. After the waxing and indigo bath, Jambi cloths were then immersed in an infusion of the wood of the *marelang* tree.[4] Although this dye was known in Jambi as *soga*, the same name as the brown characteristic of Central Javanese batik, the dyestuffs used in Java came from a different source.

In an article written by the Ethnographic Curator of the Tropical Institute, B.M. Goslings, several other differences between Jambi batik making and that of Central Java are pointed out (1927–28). For example, batik makers in Solo and Yogyakarta scraped the wax away from the cloth with a blunt knife before the blue dye bath, a practice unknown in Jambi.

Design Elements

Goslings appealed for more information from Jambi, and in response a report was received from the Jambi Controleur, Van der Kam, on current batik making practice in Jambi. The report was reproduced in Goslings' next article (1929–30). Accompanying his report, Van der Kam sent a collection of cloths back to Holland: two selendangs, two sarungs, two men's head cloths, a *kain panjang* and two pieces of rectangular cloth said to be for making up into a pair of trousers.[5] These cloths provide us with valuable data on the nature of Jambi batik, since they were collected in the field, explicitly as samples of local production. Also during the late 1920s, a Professor van Eerde obtained a

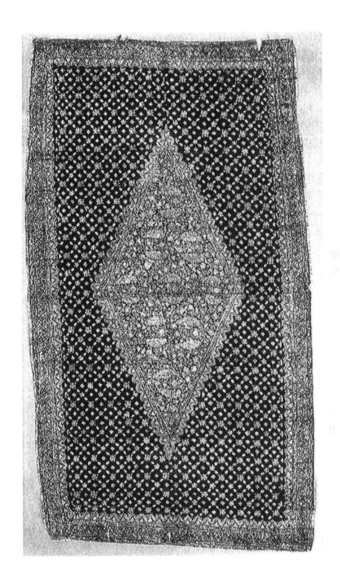

Illustration 8.1   *Selendang bersidang* (private collection)

number of Jambi batiks in yellow and blue which he gave to the Colonial Institute.[6] A collection was made a few years later by Willem Steinbuch, then Resident of Jambi, which is now kept at Museon, Den Haag. From these collections of Jambi cloths it is possible to catalogue with confidence the motifs used and the designs in which they featured in the early twentieth century in the batik making villages of Seberang.

All the cloths brought back by these collectors are hand-drawn and almost all appear to have been made with a base ground of *lembato* yellow, and an over-dye of indigo.[7] The blue areas of some have subsequently been dyed in *marelang*, making the ground area very dark blue or black; others seem to have had the major part of the ground covered in a layer of wax before the *marelang* bath. In the border areas of these, the wax layer, probably of a different mix of ingredients making the wax more brittle, has then undergone the *remukkan* (cracking) process. Immersion in the *marelang* infusion has then resulted in brown veining where the dye has penetrated the cracks. Some cloths have indigo in two shades. This has been produced by covering some of the blue areas with wax before all the indigo immersions have been completed. The appearance of all the cloths is predominantly blue or blue-black, with motifs in cream or gold, depending on how well the cloth has been protected from fading. None of the patterns has been produced by means of a stamp.[8]

Design Structures

Selendang or *kerudung*[9] can be divided into two categories: with or without a central lozenge. The majority do contain a central lozenge and are known as *selendang bersidang*.[10] Van der Kam's diagram shows another type with a small rectangular panel in the centre with the label 'sidang persegi' but there were no *selendang* of this type in the collection he sent back to the Netherlands.

Selendang bersidang have, at each short edge, a border of thin stripes, known in Jambi as *sisir*.[11] This striped border is divided from the rest of the cloth by a narrow white band which also runs along the extreme edge of the selvedge (see Illustration 8.2). There are normally three borders inside this striped edge making up a frame which surrounds the cloth: an *outer guard stripe*, the *central border*, about twice as wide, and the *inner guard stripe*, the same width as the outer. The inner guard stripe often contains the same pattern as the outer guard stripe. In simpler cloths, however, this inner guard stripe is sometimes missing. Where the borders intersect at the centre border, there is

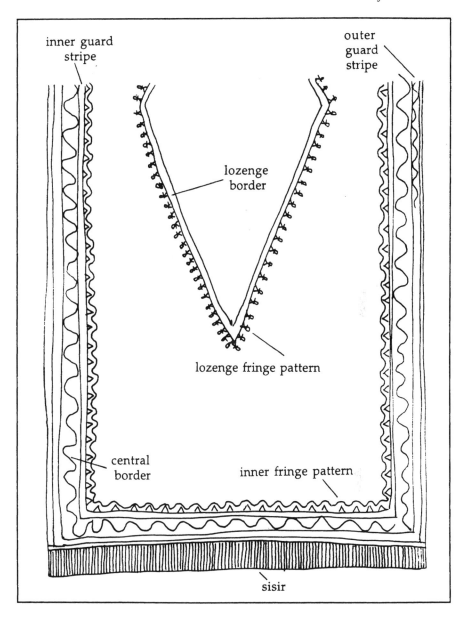

Illustration 8.2   Design structure of *selendang*

sometimes a floral or leaf motif. Along the edge of the inner border runs an *inner fringe pattern* which protrudes into the main field. This is normally described as *candi* (Buddhist temple), a word which refers to a range of motifs found in this position. It is commonly found in *songket*, gold thread brocade, from the Jambi region. Other of these patterns are dominated by large wavy lines; these are commonly referred to as *ombak-ombak* (waves).

The central lozenge is also contained within a straight-edged *lozenge border* which in turn is surrounded by an *outer lozenge fringe pattern* protruding into the centrefield. This may or may not be the same pattern as was used around the outside borders of the cloth. The lozenge border may be the same pattern as the one used in the outer or central borders, and is 2–3 cm wide. It is usually contained within a double line, but sometimes a single one. The lozenge border pattern often ends in a swollen spike.

Another type of textile produced in Jambi is *songket*, a rich silk brocade decorated with supplementary weft gold threads. The Jambi batik *sarung* is divided in exactly the same way as a *songket sarung*, into three main structural areas (see Illustration 8.3): the *badan*, the *kepala*, and the borders at top and bottom of the *sarung*. In Jambi the *kepala* is never at the end of the unsewn cloth, nor is it normally in the centre as is found in many Javanese *sarungs*. Illustration 8.4 shows a tubular *sarung* folded in half.

Men's headcloths (*iket kepala*) normally measure around a metre square. There is usually a *sisir* at the extreme outside, then a frame, again consisting of a narrow *outer guard stripe*, a wider *central border*, and a narrow *inner guard stripe* (see Illustration 8.5). The corner areas where the *sisir* end contain a foliate motif. At the intersections of the central border there is normally another foliate motif. The centrefield may contain motifs resembling those of centrefields of *sarungs* and *selendangs*.

Motifs

The classifications given to classical Javanese batik, such as *tjeplok* and *kawung* are not found in Jambi (Tirtaamidjaja, 1966). Amongst the Javanese *tjeplokkan* designs, only the pattern known as *teruntum* resembles the discrete symmetrically arranged rows of very simple Jambi designs. There are none of the contiguous or interlocking geometric patterns so characteristic of Central Java, nor do *banji* or *nitik* designs appear in Jambi. Diagonal motifs are extremely rare: the trouser lengths brought back by Van der Kam constitute the only authenticated example in Jambi batik.[12] The closest correspondence

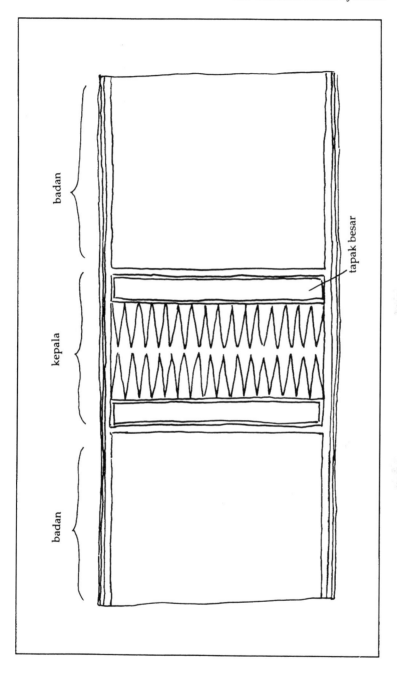

Illustration 8.3   Design structure of *sarung*

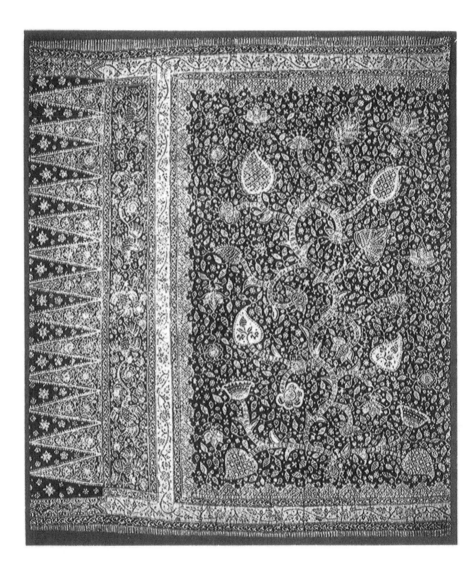

Illustration 8.4   *Sarung* in *Bataghari* design (private collection)

Illustration 8.5   **Man's headcloth with calligraphic motif in border (private collection)**

between Central Javanese batik and Jambi batik is in the category of *semen* patterns, where some of the lozenge centrefields are made up of large motifs against a background of tendrils and foliage.

Goslings' account of the design of Jambi batiks was characterized by the search for links with textiles from Java. He saw the rows of large double-motifs in the lozenge as resembling the double *lar* motif from the Javanese *sawat* pattern. However, similar shapes appear in Indian and in Middle Eastern cloths, which are just as likely to have served as models in Jambi, whose inhabitants imported textiles in large quantities from very early times (Kerlogue, 1997). However, in Jambi, where a similar motif appears in *songket*, it is known as *duren pecah* (split durian) It is more likely to derive from indigenous beliefs, in which fruits such as the durian play an important part, than from Javanese models.

Centrefield motifs are normally of rows of symmetrically arranged discrete motifs in combinations of three, four or five, often in half drops. This is also a common arrangement in *songket* design. The individual elements are usually named after flowers, and probably relate to the central use of flowers in Malay culture on ceremonial occasions, for purification, to attract the spirits of ancestors as well as to draw the community together. There are also seed and fruit motifs, which in the past were almost certainly included to evoke notions of fertility, though now they have lost these associations. Birds, associated with the upper sphere which lies between the human world and the heavenly realm, used to feature in many designs. Cloths depicting birds linked by a chain or ribbon may have been used at weddings. This motif appears in the black velvet wedding curtains of upriver districts, embroidered with silver thread. Nowadays, few bird motifs feature in Jambi batik, but two which do are the stylized *kuaw berhias* (ornamented argos pheasant) and *merak ngeram* (broody peacock) motifs. One of the most well-loved of Jambi's designs is known as Batanghari (see Figure 8.4). This is one of the large designs set against a background of leaves and tendrils. It seems to be a version of the tree-of-life, which may have been copied from Indian trade cloths or from indigenous designs. Jambi batiks in the Dutch collections feature other designs which have been lost from the Jambi repertoire.

A central lozenge appears in four other types of Indonesian cloth: the *kain simbut* from Banten, west Java; some silk *pelangi* (tied or sewn resist) cloths, usually catalogued as coming from Palembang, though they were also made in the Jambi region; the Javanese *dodot* and the *kemben*, common in Java but also found in Bali. However, only in the *kain simbut* and the *pelangi* cloths are the dimensions and arrangement of the cloth similar to those of the Jambi *selendang bersidang*. There is also a striking similarity between the design arrangement of these lozenge *selendangs* and certain carpets and *kilims* originating in the Middle East. *Kilims* and other items brought back from the Haj, the pilgrimage to Mecca, are found in many houses in the village, and are usually regarded as having supernatural protective properties. It is not at all unlikely that designs might have been copied from these precious items. Certainly there is a strong Islamic aesthetic at play in the design of the cloths, with some containing calligraphic motifs derived from the Qur'an (see Figure 8.5).

The range of batiks produced in Jambi today is far wider than it was in Adam's day. Although contemporary batik makers still use many of the old designs they have now extended their repertoire. In addition, it now seems clear that in the nineteenth century Jambi produced a range of other cloths, some block-printed and some containing the rich red colour produced by dyeing with the root of the *mengkudu* tree. Another group of textiles where the two faces of the cloth are decorated with different patterns was also produced in Jambi, and one batik maker there today still produces cloths of this type. However, any collector, curator or scholar finding a cloth with the elements found in the Adam, Van der Kam and Van Eerde cloths outlined in this article can be fairly certain that it was produced in the classic period of batik making between the mid-nineteenth century and the 1930s, in one of the villages of Seberang.

Notes

1   See, for example, McCabe Elliott (1984).
2   Now kept in the Tropical Museum, Amsterdam and catalogued as no 347/2.
3   Identified as *Artocarpus cf. dadah* (*Moraceae*) by the Royal Botanic Garden, Edinburgh. It may be the same as *Artocarpus Limpato Miq.* (*Moraceae*) referred to in Uphof (1968) as being native to Sumatra.
4   Identified as *Pterospermum cfg acerfolium* (*Sterculiacae*) by the Royal Botanic Garden, Edinburgh.
5   These cloths are in the Tropical Museum, Amsterdam, catalogued as nos. 483/1 to 483/8. No. 483/6 appears, however, to be missing.
6   The cloths are catalogued as 556–5 to 556–18.
7   Some cloths containing red were also sent back to Holland. The dispute surrounding their origin is explored in Kerlogue, 1999 (forthcoming).
8   The absence of stamped batiks in these collections does not preclude the possibility that such a technique existed in Jambi. Hand-drawn batik has always been more prestigious than stamped batik, and it is likely that Van der Kam was provided with cloths of the best quality. See Kerlogue, 1999 (forthcoming).

9   In relation to pre-independence textiles, these terms are used interchangeably. The term *kerudung*, however, is not used to refer to the recently-introduced narrow Javanese-style *selendang*, which is not worn as a headcovering but over the shoulder for formal occasions.

10   See Geirnaert-Martin (1983) concerning the significance of lozenges in Javanese batik.

11   Goslings described this as a '*pengada*' edge; 'a completely Javanese characteristic', by which he must have meant that it is also found in some Javanese batik. It is probably more characteristic of Jambi, in that it appears in all *selendang bersidang*.

12   They may have been made especially for the Europeans: batik trousers were often worn by Dutch colonial officials at home.

References

Elliott, I. McCabe (1984), *Batik: Fabled Cloth of Java*, New York, Clarkson N. Potter.

Geirnaert-Martin, D. (1983), 'Ask Lurik Why Batik: A structural analysis of textiles and classifications (Central Java)' in J.G. Oosten and A. de Ruijter (eds), *The Future of Structuralism: Papers of IUAES Intercongress, Amsterdam 1981*, Amsterdam, Gottingen.

Goslings, B.M. (1927/8), 'Een batik van Djambi', *Nederlandsch-Indie Oud en Nieuw*, 12, pp. 278–83.

Goslings, B.M. (1929/30), 'Het batikken het gebied der hoofdplaats Djambi', *Nederlandsch-Indie Oud en Nieuw*, 14, pp. 141–52.

Heringa, R. (1994), *Een Schitterende Geschiedenis*, Den Haag, Museon.

Jasper, J.E. and Pirngadie, M. (1916), *De Batikkunst. De Inlandsche Kunst-Nijverheid in Nederlandsh-Indie, Vol III*, S'Gravenhage, Mouton.

Kerlogue, F. (1997), 'The Early English Textile Trade in South East Asia', *Textile History*, 28, No. 2.

Kerlogue, F. (1999, forthcoming), 'The Red Batiks of Jambi: Questions of provenance', *Textile Museum Journal 1997/8*, Washington DC, The Museum.

Raffles, T.S.B. (1817), *The History of Java*, London, Black, Parbury and Allen.

Tirtaamidjaja, N. (1966), *Batik: Pola & Tjorak*, Jakarta, Penerbit Djambatan.

Uphof, J.C.T. (1968), *Dictionary of Economic Plants*, New York, Cramer.

9 The Development of Motifs in Indonesian Batik Fashion Trends from 1850 Onwards

ARDIYANTO PRANATA

Introduction

Javanese batik has emerged as a rare and exotic art form over the past 150 years, although its origins are ancient and widespread. Java is a trade route island. For hundreds of years the Javanese coast has lain at a crossroads of trade, near the course sailed by Marco Polo, Magellan, Sir Francis Drake and St Francis Xavier. Over the last 2,000 years Java has been influenced by a succession of religions – Hinduism, Buddhism and Islam – and waves of colonization – by Chinese, Indians, Arabs, Portuguese, Dutch and English. Each of these influences left its imprint on Javanese culture and the development of batik.

The influence of the Chinese on Javanese batik was as profound as that of the Muslims, the Buddhists and the Hindus. They brought mythical lions and lyrical flowers to batik design along with a bright new palette of colours.

The diversity of Java, spurred by its growing population and the freewheeling mercantile life of the north coast, spawned an ethnic group known as *Indische*. They were part Dutch and part Javanese, but they could also be Chinese or Arab (combined with any European nationality), or part Chinese, part Javanese, or even pure Dutch who had lived for generations in Java and thus were 'native' to the Indies. The Indische came to be influential designers of commercial batik (but not the biggest entrepreneurs; that distinction belonged to the Javanese). The Indische group was highly influential and productive for about 70 years in Pekalongan and in other northern towns of Java.

Batik history, motifs, design and methods of production are closely associated with their origin. Thus, as one travels from the provincial and stately courts of Yogyakarta to Cirebon, Pekalongan, Lasem and other towns on the commercial north coast, the diversity of batik is readily observable.

The modern design era commenced with the breakthrough in traditional motifs that was accomplished by distinguished Indonesian batik artists. The modern trend varied from the unification of both north coast and central Javanese motifs and dramatic enlargement of traditional patterns (Harjonagoro); the elaboration of North Coast batik in wall hangings and the simplification of usually saturated design and combination of tie dye with hand printed batik on the same fabric (Ardiyanto Pranata); innovative use of materials as well as the unusual large and bold designs (Iwan Tirta); until the use of specially designed hand-woven fabrics (Obin and Baron).

Cirebon Batik

The heavy hand of Dutch colonialism radically altered the way batik was made. In 1830 the Dutch imposed heavy taxes on agricultural acreage. This meant that men had less time to make batik and as women slowly replaced men as makers of batik, the bold, masculine approach to design began to fade away as seen in Cirebon batik.

In the course of their rule the Dutch encouraged the rise of a Chinese entrepreneurial and bureaucratic class; they even leased whole villages to Chinese who collected taxes, sold opium and controlled the labour force. Chinese communities sprang up – Kanduran was one – and their citizens settled in and married local people, they adopted the local style of dress, retaining their own patterns and colour preferences. Chinese artifacts were well known there: the *kylin*, a figure that was half dog, half lion; the *banji*, a swastika-like fretwork motif; and more common designs, such as the phoenix, peony and chrysanthemum. These Chinese motifs were eventually incorporated into the religious batik of Cirebon.

As long as 100 years ago, these Chinese Cirebon batiks had become popular with the Europeans, both as collectors items and as clothing for personal use. The new styles, however, were almost never worn by the local people: the multicoloured patterns were quite alien to them.

Pekalongan: The City of Batik

In contrast to courtly central Java and Cirebon, batik in these coastal towns was not so much made for family or personal use; trade was the name of the game. Batik factories, some employing hundreds of workers, were established by entrepreneurs who catered to local and overseas tastes.

In the 1920s, the 'golden era' of batik production, thousands of people were employed in the various coastal centres. As the industry grew, the demand for skilled workers increased and hordes of people swarmed from town to town selling their labour. But the finest batik was made in smaller factories, usually by people of mixed descent.

Java's north coast batik illustrates much of the area's past and present. The batik is surprising and exuberant, vigorous and muted. It explodes with brilliant reds, traditional blues and radiant yellows blending with soft pastel tints of green, lilac and pink. The motifs – people, farm produce, fish and animals – evoke the coast of Java in all its varied life.

In the first half of the nineteenth century the Dutch had exported imitation printed batik to Java, using cotton from the United States. The American Civil War stopped this traffic in the 1860s, just as an expanding Dutch colonial administration was creating a need for more cloth. To meet this demand, Chinese, Arab, Javanese and even some Dutch factories appeared and batik-making began to progress from its cottage industry beginnings to larger workshops. This is what happened in Pekalongan and why the town ultimately became Batik City.

European Batik

Among the Indische entrepreneurs were women such as Mrs Simonet (her only known name), Lien Metzelaar, S.E. Bower, J. Jans (also known as widow Jans), M. de Ruyter, P.A. Toorop and Eliza van Zuylen. In a town as small as Pekalongan, they must all have known one another and they also knew Arab traders and other Chinese batik entrepreneurs. Among them they created many memorable batiks.

Mrs Simonet was Chinese-Indonesian, married to a Frenchman. She had a preference for romantic illustration, exemplified by a batik of flying cupids on a green background. Mrs Simonet's daughter took on the atelier of Lien Metzelaar, who had worked in Pekalongan from about 1880 to 1920. This atelier was known for batik with a border of seven leaves on a straight branch alternating with four simple flowers; Metzelaar batik is easy to recognize from this motif.

The workshop of J. Jans catered to the wealthy at the turn of the century. An early Jans sarong shows the 'lace border' with delicate red scallops set against a clean, creamy background of repeated stars and flowers. The motifs changed when art deco was introduced in the early 1900s; stylized irises and swans in a lily patch all combined with the typical Chinese swastika or *banji* in the background. Jans sometimes added ground betel nut and chalk to the natural dyes to soften her colours.

Eliza van Zuylen was another fine example of the Indische entrepreneur. Her life spanned nearly a century, from 1869 to 1947, coinciding with the heyday of Pekalongan batik and fortunately her life and work are well documented. Born in Jakarta, she moved to Pekalongan where she established a batik compound in 1890 at what is now a police barrack. As in other batik factories, van Zuylen's whole family was involved, with the women in charge of design and personnel and the men controlling the dying and fiscal matters. These work patterns persist in Java's batik factories today.

Chinese Batik

By the time Pekalongan began to launch its booming batik business in the 1850s, the most significant designers were The Tie Seit, Oey Soen King, Oey kok Sing, Liem Siok Hien, Lim Boe In and Liem Boen Gan; in Kedungwuni, Oey Soe Tjoen was the most important.

In Chinese custom age had a lot to do with who wore what. Light pinks and blues were worn by younger girls, blue and red by middle-aged women, while the elderly wore a combination of blue, brown, purple and green on an off-white background. If the father in a family died, his wife, children and grandchildren traditionally wore blue and white sarongs for two years as a sign of mourning. Red symbolized good luck, and if a baby was expected, a gift of a red carryall or *selendang* was presented to the parents. Especially if the family was rich, a *selendang* decorated with gold would be ordered to celebrate the birth of a first-born son.

Beautifully-made batik *tulis* was often part of a bride's dowry. There was a special batik, the Cempakamulya, made to be worn by the parents of both bride and groom at the wedding ceremony. A week after the wedding, the bride was presented with her own Cempakamulya (*Cempaka* is a gardenia and *mulya* means 'virgin'). The circular flower of this design has a pointed petal often combined with a diagonal *parang menang* in the background, symbolizing wisdom. Sometimes the pattern includes a curved continuous vine, signifying long life.

Batik made by the Pekalongan Chinese has two distinctive appearances.

Pre-1910 Chinese batik resembles traditional north coast batik: the rich natural colours of indigo blue and *mengkudu* red are set on cream and tan grounds. Usually these pre-1910 textiles were made at home, for family or ceremonial use. Motifs were drawn from Chinese art and applied to the traditional layout of the *kain panjang*. In the town of Lasem the Chinese made similar batik and it is often difficult to distinguish between those of Lasem and of Pekalongan.

In these pre-1910 batiks, the head (or *kepala*) of the sarong was usually split: red on one half and black (red over-dyed with indigo blue) on the other. An elongated triangle shape or *tumpal* was set against red and blue medallion motifs, edged top and bottom by borders that seem to evoke India more than Java or China. Often the body of the cloth contained diagonal stripes, snake motifs or phoenix-like birds and lotus flowers, reflecting its Chinese origins.

After 1910, Chinese Pekalongan batik burst into full colour with a panoply of flowers, multicoloured shadings and tiny textures of intricate fillings (*isen*) for both the designs themselves and the backgrounds. These artistic explosions coincided with the introduction of synthetic dyes which the Chinese used much earlier than the Indische.

During this period, European designs became a symbol of rank, a manifestation of the Chinese ascent in society; but as always with batik, motifs melded. The Dutch tulip, for instance, was soon transformed into the Chinese lotus.

The batik of Pekalongan's The Tie Siet clearly shows his Chinese heritage. He brilliantly used orange and royal blue, sometimes aqua, weaving these non-traditional colours into such age-old motifs as peacocks and birds. Although The made sure to fill each leaf with a multitude of detail, the backgrounds were usually plain with vivid art deco borders. In a carnival of batik, the work of The Tie Siet is always recognizable. And like most of the Chinese batik made after 1910, they were for sale and not for family use.

In one Pekalongan house, the women of the family have been making batik for three generations. The grandmother, Oey Soen King (née Liem Loan Nio), was born in 1861. Cloud motifs, serpents, dragons, lotus flowers and bats, a sign of good fortune, were characteristic of her batik. Although she lived until 1942, she always used natural red and blue dyes in classic layouts of the late nineteenth century. Oey Soen King's pieces were all unsigned, since signatures were not yet in vogue.

Her daughter, Oey Kok Sing (nee Kho Tjing Nio) was born in about 1896 and died in 1966. Oey Kok Sing chose a different palette and new motifs but displayed the same control and grace as her mother. She used vibrant synthetic colours to enliven her 'European' bouquets. A secondary patterning characterizes the work of Oey Kok Sing: the cloth is packed with a background of intricate work that seems to echo the foreground motifs. Oey Kok Sing left many cloths to which she had signed her name.

Oey Kok Sing's own daughter, Liem Siok Hien, whose maiden name was Oey Djien Nio, was born in 1929. Not satisfied with the typical Dutch bouquet, she chose to lace flowers through her intricate batik paintings, introducing new combinations of exotic birds, rock formations and mythical animals set on clean, cream ground. Her soft backgrounds always have a surprise element: tiny castles and kernels of rice add texture and shape to the batik. The tulis work is so fine that a *kain panjang* often took Liem Siok Hien 14 months to complete. She always mixed her own colours, achieving a spectrum not found in the batik of her mother or grandmother. In 1965 Liem Siok Hien began signing her Indonesian name, Jane Hendro Martono.

Oey Soe Tjoen, who died in 1975 at the age of 74, was known for his craftsmanship, his attention to detail and his attention to the taste of his clientele. Unlike van Zuylen, Oey Soe Tjoen did not use plain pastel colours; his family did not think them appropriate for their largely Chinese clientele.

Oey Soe Tjoen numbered many of his patterns: it was easier for his customers to order that way. He also signed all his pieces, usually making sure to include the name of his village Kedungwuni.

What and who inspired the design of Oey Soe Tjoen? Some designs came from Pekalongan where the family would buy artwork and then adapt the patterns and colours to suit their clients. Others come from a book of flowers – a gift from a Dutch dye company. These floral motifs were usually brightly re-coloured in typical Chinese fashion.

Although most of these Oey's early customers were Chinese, an increasing number of Javanese motifs were no doubt incorporated at a client's request but there may have been other sources of inspiration as well. One of these was perhaps Princess A.A. Ario Soerio (b. 1901) who took it upon herself to 'teach' the north coast batikers something about 'real batik'. An energetic lady, she travelled to northern Java accompanied by the best batikers from the royal courts of Yogyakarta and Surakarta, demonstrating how to make isen (textured filling) instead of the usual dots, how to incorporate the royal crest and how to make *soga* brown. There is a certain irony in a noblewoman from central Java travelling north to teach about *isen*, royal crests and *soga* brown since these techniques, dyes and motifs all probably originated on the north coast.

By Chinese standards at least, the Oey family of Kedungwuni was late in switching to synthetic dyes, which they did in 1928, for each new colour the cloth was immersed in a dye bath, unlike the practice of some Pekalongan workshops, colours was never painted on the cloth even for the tiniest details. Prior to 1942, as in the van Zuylen factory, Tjap Sen cloth imported from Holland was used for batiking.

After World War II the Oeys continued their superb craftsmanship as before; sales were generally left to middlemen and the design changed little except by customers' request. To this day the batik of Oey Soen Tjoen is considered the very finest.

Silk Batik

Batik on silk, like batik covered with gold (*perada*), is a sign of wealth and prestige. Probably no batik is more significant than a silk that has been dyed yellow to be used as a religious banner or in a shadow puppet play. Yellow is the colour of royalty. A yellow carryall (*selendang*) from Juana, when wrapped around a puppet in a *wayang* performance, represents Batara Guru, the Father of All, Java's most important deity.

Silk batik comes in many other colours and shapes. One type, called *lokcan*, was made in Juana in the 1920 by Tan Kian Poen who patterned the silk with small flowers and animals in reds, blues, browns and yellows. These silk (*can*) batik pieces were often dominated by a shade of blue (*lok*), hence the name *lokcan*. Unfortunately fine silk batik is rare and most of it has long since succumbed to Southeast Asia's hot and humid climate.

Dua Negeri and *Tiga Negeri* Batik

Several batik forms are referred to as 'two countries' (*dua negeri*) and 'three countries' (*tiga negeri*) and these are combined in the best styles and colours from several Javanese batik-making towns, all of which participated in their production. *Dua* and *tiga negeri* batik spread along the northern coast of Java, extending east to Surabaya and south to the central highlands. Motifs and colours of *dua negeri* and *tiga negeri* batik are usually so distinctive that the origin of each piece can be easily traced. Usually the first waxing and dyeing was done in north coast towns; the second waxing and final dyeing in central Java. 'Three countries' batik was expensive because it travelled farthest and was thought to combine the best of all Javanese styles, making the 'perfect' batik.

A *dua negeri* sarong, for example, might originate in Pekalongan where the head (*kepala*) and border (*pinggir*) were waxed and dyed red before the cloth was sent on to Surakarta, where the body (*badan*) would receive *garuda* wings and other central Javanese motifs before being dyed *soga* brown. A 'three country' batik might originate in Lasem where pomegranates and vines, the main designs, were waxed and dyed red; it then might go to Kudus for these motifs to be filled and dyed blue and it might end in Yogyakarta were a *parang* motif would be added to the background and the final colour, *soga* brown. Styles and dyes in these batiks were almost infinite. *Dua* and *tiga negeri* batik was made for about 50 years, until World War II effectively halted its production.

Hokokai Batik

Hokokai batik was made all along the north coast, especially in the Pekalongan area. While the flowers retained their regional styles, most towns used similar background and colours making it difficult to ascertain exactly who made which batik; few *Hokokai* pieces are signed or stamped.

Rarely does one encounter a *Hokokai* sarong. Instead, morning/evening pieces (*pagi-sore*), which use more cloth than either a sarong or a *kain panjang*, were produced. Given the cotton shortage, this was surprising: it may be that *Hokokai* was commissioned on *pagi-sores* because the Japanese thought that batik could best be displayed on such lengths; or perhaps the Japanese used *pagi-sores* for other unknown purposes. *Obis* – the sashes tied around kimonos – were also batiked during this period.

Visually *Hokokai* was a natural follower of some of the Chinese-inspired batik. The bright colours remained, but the combinations changed during the war. No other era in batik history is as specially delineated and as brief as *Hokokai*. But as with many other things in the world of style and design, *Hokokai* batik was revived in later years.

Indonesian Batik Breakthrough

Mrs Bintang Sudibyo, better known as Ibu Sud, was the first designer who achieved great recognition. She designed a *kain panjang* with new variations on the already-existing Principalities patterns and the patterns that she borrowed from the Hindu-Javanese temple reliefs. For this, she made use of a colour scheme that she borrowed from the north coast batiks. After a visit to Thailand, she assimilated classical Thai motifs into her batiks and in 1969, she introduced the Indonesian variant on the Thai dress, made of batik with Thai modified motifs. Every year, Ibu Sud sought to bring out onto the market a new collection of clothing with a new fashion image and new batik designs.

Another innovator was K.R.T. Hardjonagoro, who earned a lot of success with his batiks. Hardjonagoro's ability to unify both north coast and Central Javanese designs, as well as his dramatic enlargement of traditional patterns, was intriguing.

He had studied with Mrs Bintang Sudibyo, who had been the instrument in developing indigenous Indonesian taste by using green and purple rather than the colours preferred by the Europeans and Chinese. Given that he was a connoisseur of the Solonese court culture, he went on to create batiks in the specific Solonese style. These designs are not intended to be inspired by fashion, but instead are based on the Javanese philosophy, and from a technical standpoint, his batiks belong to the best which are presently produced in the Principalities. Next to the batik *cantings*, he also produced more light-hearted designs with the *canting* and *cap*. With these, he shows himself to be more playful and also innovative.

Pekalongan's Achmad Yahya is another leading maker of modern-day batik. Yahya and his Cirebon-born wife exemplify how batikers have adapted to consumer demand, rather than slavishly following old influences. He had produced only sarongs and *kain panjangs*, which were sold locally. Within two years he had learned to make batik by the yard, with complicated motifs using both *tulis* and *cap*. New design configurations were drawn and special tables for waxing were erected. After mastering batik by the yard, he then learned about dyeing so that he could achieve a pastel palette on yard goods.

Achmad Said is a truly original batik artist and manufacturer whose family came to Surabaya in the fifteenth century. The workmanship of Said is not especially distinguished – he makes only *cap* batik – but it is the wildly extravagant patterns that make his work distinctive. Working on primissima, a fine cotton ground, Said splashes his sarongs and *kain panjang* with a bold array of geometric background on which bouquets of flowers are superimposed, creating a two- or three-layered effect. No-one else has put geometry, nature and colour to such imaginative use.

Thirty-eight miles to the east of Cirebon, H. Mohammed Masina and his wife have revived some of the secret royal patterns of the court of Cirebon. It was not the Masinas, but another couple, the Madmils, who first unearthed some of the noble Cirebon designs. Masina and Madmil began producing stocks inspired by old designs for the clients from Jakarta. A simple execution of a sarong with the van Zuylen bouquet and an accompanying *selendang* were made by Masina and became the rage at the end of the fasting month in 1979. Many women came in chauffeured cars from Jakarta to Trusmi (a trip that surely took five hours) to do their Lebaran shopping. The van Zuylen bouquet enjoyed, albeit for a short period, a comeback. The customers no longer associated the van Zuylen bouquet with the colonial past, but viewed it as a new fashion. Because the batik work of the Masinas is so extraordinarily fine, they are given the credit for this renaissance. Iwan Tirta, too, was important in this development, because he popularized the large royal lions on oversized cotton sheets, sometimes even commissioning the Masinas to batik his designs.

Both Masinas are third generation batik makers who use traditional browns on cream or tan grounds. Chinese blues and reds characterize their palette as well. There are few north coast batikers these days who produce Cirebon batiks in *tulis* as well as the Masinas. Indeed, had it not been for the Masinas' fine technique, there would have not been anyone to carry on the Cirebon batik tradition.

After 1960, a new group of customers appeared in Java who could afford the relatively expensive batik *canting*. This included the women from the new elite, those who occupied high positions in the government. Batik designers sought inspiration in the past to fulfil the increasing demand few new batik patterns. There were many changes in batik during this period.

Nursjiwan Tiraamidjaja, better known as Iwan Tirta, after completing his studies in law, decided to pursue a career in batik. He sought antique batik Cirebon and found these among other in Dutch museum collections. He had them copied in Trusmi, the batik centre outside of Cirebon and he also made new designs in the traditional style.

Cloth for batiking was, traditionally, at most 40 inches wide. Not until the late 1960s did batik motifs explode, when Jakarta's Iwan Tirta began to use cotton that was 58 inches wide. Tirta went one step further: he used these wide batiks with their massive figures of dragons and lions for women's clothing. It was a break in tradition, for while large motifs had been worn by men in Javanese high society prior to World War II, women were always more modestly clad in *kain panjangs* of smaller, more delicate designs.

Tirta's batik attracted a great deal of attention in the international fashion world because of his innovative use of materials, including silk, and his unusual designs. Although few batik workers today are capable of fine work, they can quite easily wax Tirta's large, bold designs. Tirta pioneered in several ways.

He enlarged traditional patterns with a few bold colours; he made batik for both clothing and home furnishings; and he experimented with *cap* batik by the yard.

Iwan Tirta, the best known fashion designer in Indonesia, earned the reputation by introducing new batik designs for costumes, such as a silk *kain panjang* accompanied by a very large *selendang* batiked on silk. They were inspired by the silk batik *selendangs* and sarongs that were made by the Chinese batik manufacturers in Juana and Rembang until around 1920 for their Chinese customers – flowers and a phoenix in brown-black on an ecru or beige background.

Among Indonesian batik artists, Obin is one of the most creative and innovative, which is imprinted in her work. A specially designed hand-woven fabric is used as the raw material for her batik. Baron, on the other hand, uses a very find hand-woven silk for his batik.

As one of Indonesia's batik artists, I contributed in developing further the traditional batik design through elaboration on north coast batik in my wall hangings. Since 1994 I have experimented in combining batik techniques with tie dye on one fabric. However, this is not new, since this method was used before the second world war. This technique requires experience for optimization.

In the four decades since 1945 batik has become more Indonesian, even international, than strictly Javanese. It is now almost impossible to recognize the origin of modern motifs, much less the designer. What is interesting about these modern trends and designers is that their distinctive styles are adapted from all parts of Java, especially the north coast. Their roots are as diverse as their batik.

Around 1970, the Governor of Jakarta, Ali Sadikin, decided to stimulate the batik industry by introducing the batiked men's shirt with long sleeves as the formal attire for Indonesian men. This initiative was generally accepted, due in part to his great popularity in all levels of society. Again, we see that the designers took the initiative by introducing observable fashion trends in the batik designs for men's shirts. In the beginning, the shirts from Ibu Sud were popular and after that the shirts from Iwan Tirta were a must for well-off customers.

Around 1980, the batik print appeared. This has nothing to do with the batik technique. It is a one-sided print in which the diverse colours are applied to the fabric via screen-printing. The batik print became so successful that together with the batik *cap* they seemed to replace the batik *canting*. Many customers did not know the difference and did not really care, either. And the

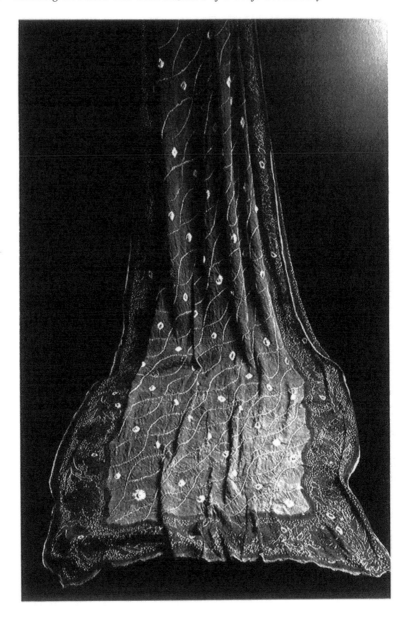

Illustration 9.1a   Ardiyanto Pranata: batik and tie dye silk shawl, 2.2 x 1.1 m, 1998

**Illustration 9.1b   Ardiyanto Pranata: batik and tie dye silk shawl,
2.2 x 1.1 m, 1998**

tourists were also more interested in the price rather than in the technique. For them, the mention of 'batik' was more than satisfactory.

The future of the batik *canting* industry looks sombre. But there is talk of a turnaround in favour of batik *canting*, albeit on a limited scale. For weddings, on which relatively much money is spent in Indonesia (as elsewhere), the batik *canting* has become again the favourite for confirming the social position of the families involved, not only for the family of the bridal couple, but also for the guests. At weddings, people strive to dress 'in the costume of their forefathers' and so far as that concerns batik, we can assume that this can only be for the good of the batik *canting* industry.

Presently, many of the invited male guests wear a European suit to the wedding receptions and those women who can afford it mostly wear batik creations from the great designers, while the Chinese visitors of the older generation, in many cases, just as in the past, still wear the sarong with the van Zuylen bouquet. For them, this is still the traditional costume.

10 The *Garuda* Motif and Cultural Identity

WORO ARYANDINI S.

There are various motifs in batik cloth, ranging from vegetation to animals and even to human figures. This chapter will deal with the bird motif.

The bird is either illustrated wholly or sometimes only as wings. In one instance, both wings are depicted; in another, only one of the wings is shown. In batik terminology the decorative element of wing is called *lar* or *elar*. Sometimes the motif consists of both wings with a tail, *sawat*. Alternatively, the *lar* is combined with the *semen* motif in various forms. '*Semen*' comes from the word '*semi*', meaning 'the sprouting of new buds'. Birds and trees are in fact closely related since birds live in trees.

Batik and other arts and crafts designers often incorporate certain decorative elements without the complete understanding any longer of the original philosophic meaning of the decorative elements. According to Van der Hoop (1949, p. 10), there are decorative elements in Indonesia which resemble motifs of Hindu-Indian origin. Nevertheless, the bird motif may pre-date contact with India since it is encountered in Indonesian cultures that lay beyond Indic influence. Indic motifs may also have been accepted by the Indonesians, because these were already identified in the original Indonesian culture. The adoption of the bird as a symbol or a decorative motif possibly combines Indic and pre-Indic beliefs.

Of all birds, *garuda* is of special importance. In Hindu mythology, the *garuda* is the vehicle ridden by Vishnu, the god who maintains the universe. The *garuda* motif is often found in Hindu-Javan arts, sometimes with Vishnu, but quite often by itself. *Garuda* is also the symbol of the sun and is called the sun-eagle, symbolizing the world above, the opposite of the serpent, the symbol of the world below. Both are used as batik motifs.

Garuda motif is also important to the Indonesian people as national emblem, the *Garuda Pancasila*. It is named *Garuda Pancasila* because the bird's chest is in the form of a shield, composed of *Pancasila* symbols, the philosophy and basis of the Republic of Indonesia. The *garuda* symbol in slightly different forms is also used as the emblem of: the Department of Education and Culture; and the Indonesian airline company, Garuda Indonesia Airways. In the past, in the eleventh century, *garuda* was used as the stamp and the logo of the royal court of Kahuripan in East Java. The royal stamp was called *Garuda-mukha.* and its king, Erlangga, was depicted as Vishnu, riding on *garuda*. Regarded as being so important, the *garuda* motif is used in KORPRI (Indonesian Civil Servants Corporation) uniform. Here, the two wings of the bird is used in the KORPRI batik motif with a house and a wishing tree above.

The eagle is widely used as a national symbol (e.g. Austria, Germany) and is usually depicted as a stylized version of the natural bird. Live eagles are also sometimes kept in national showcase aviaries, as is the case in Taman Mini in Indonesia. The mythical *garuda*, however, turns out to be different from the ordinary eagle-type bird. Garuda's story is told in the Ādiparwa, the opening of the Mahabharata epic which originated in India. The story partly explains why Garuda was chosen to represent the Republic of Indonisia. The story goes like this.

There were two wives of Bagawan Kasyapa, named Kadru and Winata. Both of them had long been married but none had borne a child, so they appeared before Bagawan Kasyapa. Kadru requested to be given 1,000 children while Winata asked for only two children but who would be supernaturally more powerful than Kadru's children. Bagawan Kasyapa gave 1,000 eggs to Kadru and two to Winata. After 500 years, Kadru's eggs hatched into snake-form children, among others Anantaboga, Basuki, Taksaka, who all had great supernatural power. Winata's eggs had not hatched by then, so she felt ashamed and defeated by Kadru. She cracked one of the eggs. The baby looked like a bird, but was still premature with only the upper body fully formed. The face was perfect, but both legs were still missing. The baby cried because it had been hatched prematurely, causing its body to be deformed. He said to his mother: 'Mother, I feel so painful because my body has not grown fully. You had wished your children to be superior to Kadru's children. One day you will be enslaved by Kadru, and you will suffer a lot. Be careful in looking after my only brother, because he is the one who would set you free from slavery.'

The child was named Aruna, meaning limbless; he could only sit. He later was appointed as coachman to Sanghyang Aditya or Sanghyang Surya. Every morning, before the sun appears in the east, we see a red beam of light: it is Aruna.

One day the gods wanted to obtain *amrta*, the water of immortality. Vishnu informed them that in order to find it, they must stir Ksîrârnawa, the Milk Ocean. After long and tiresome stirring, they found *amrta*, which emerged together with some creatures, one of which was a horse named Uccaihscrawa. Kadru told Winata that the horse was white with black spots on its tail, while Winata reckoned that it was all white including the tail. Each defended her opinion and in the end they agreed that whoever was wrong would become the slave of the other. They would together see the horse the following day.

Kadru told her sons that she was about her bet with Winata, but the sons said that Winata was right and that the horse was all white. Kadru was confused because she was going to lose and become a slave to Winata. She then tried to persuade her sons to spray the horse's tail with black spots with their poison saliva. The sons refused Kadru's request, not wishing to disgrace themselves. Full of indignation, she threatened to curse her sons to become the victims of snake-offering fire held by Maharaja Janamejaya, if they did not obey her instructions. Finally the dragon sons obeyed their mother and sprayed the horse so that it had black spots. Winata lost the bet and became Kadru's slave.

As Winata was living as a slave, her second egg hatched and a vigorous and handsome bird emerged. Not finding its mother, it flew to the sky. The feathers were glittering with lights like a fire illuminating the universe. The gods saw this with awe, thinking that the time of the destruction of the world had finally come. Batara Agni exclaimed: 'Behold, gods, do not fear, it is not yet time for me to destroy Tribuwana, the time of the destruction has still long to wait'. 'What is then that shiny thing?' asked the gods. 'That is the beam of Sang Garuda, the mighty and all powerful bird, the son of Bagawan Kasyapa and Sang Winata.' All of the gods praised Sang Garuda: 'Lord Garuda, you ascetic, you the divine one, you god, who shines like Sanghyang Surya. Protect us, because you are the most perfect bird'. So went the gods' praises, and made Garuda happy.

He then arrived at the place where his mother served as a slave. He was asked to stay there to serve Sang Kadru, watching the dragon-sons. The dragons always roamed around and made him tired, let alone his mother. He asked his mother why she should obey Sang Kadru's orders, which in turn become his tasks as well. Winata explained to him that she was currently a slave to Sang Kadru, 'If you take pity on your mother, go and ask Sang Kadru and the dragons, what would compensate for my freedom'. They replied that they wanted *amrta*, which was in the possession of the gods. Garuda was happy to hear this. He said goodbye to his mother and asked her for advice in searching for *amrta*. She advised him first of all to eliminate the wicked, and further said that when he devoured his victims and felt heat in the throat, it meant that he was swallowing a holy man. He should put him out. 'Those holy men are like your father, Bagawan Kasyapa, whom you have to respect. Go, my son, Hyang Bayu will protect both of your wings, Hyang Chandra your back, Hyang Agni your head, all gods your entire body,' so said Winata.

He soon flew to the location of *amrta*. The gods guarding the *amrta* shot him with weapons but all these were broken and did not hit a single feather. Garuda retaliated by pecking his opponent's eyes, blinding them. *Amrta* turned out to be surrounded by ever-burning fire. Garuda was not short of ideas; he gulped a lot sea water and sprayed it onto the fire until it was extinguished. He saw *amrta* inside a cave, guarded by revolving *cakra* (spiky weapons) at the gate. He shrank himself so small that he managed to pass through the *cakra*. Arriving at the pot containing *amrta*, he was confronted by two dragons with unblinking eyes. Sang Garuda launched himself toward the dragons while flapping his wings rigorously, throwing dust into their eyes and blinding them. At last he succeeded in getting *amrta*.

On the way home Vishnu stopped him and requested him to hand over the *amrta*. At the same time Vishnu asked whether or not Garuda was willing to become his vehicle. *Garuda* briefly considered the offer and replied that he was willing to become Vishnu's carrier. Then Batara Indra came and asked Garuda to return the *amrta* to the gods. Garuda said: 'The gods may use *amrta* to their heart's content after my mother is free.' He then flew back to his dragons. He tied the *amrta* pot with *ilalang* (long, coarse grass; *Imperata cylindrica*); his mother was freed.

When giving the pot to the dragons, Garuda advised the dragons to purify themselves before drinking *amrta*, by way of ablution. The dragons were so busy struggling to be washed first, that the *amrta* was left unguarded. Seizing the opportunity, the gods took back the pot. The dragons, upon leaving the ablution area, found that the *amrta* pot had gone; there was only a drop of *amrta* on the *ilalang* blades. They struggled and each tried to drink a little of the drop left on the *ilalang*. Out of the struggle, their tongues were cut into two by the sharp edges of the *ilalang* blades. Having absorbed the immortal water, the grass still lives. So goes the story of Sang Garuda.

This story is carved on the mirror handle which is preserved today at the National Museum in Jakarta. On the front side of the handle, Garuda is shown saying goodbye to Winata, his mother, before going in search of *amrta*. On the back, Garuda is depicted at the scene of his escape from the gods, carrying the *amrta* pot against his chest. Fragments of the story of Garuda (Garudeya) are also found on the relief of *candi* (temple) Kidal (thirteenth century), *candi* Kedaton (c. 1370), both in East Java. At Candi Sukuh (fifteenth century) on the slope of Mount Lawu, the relief depicting Garuda's story is more complete. Garuda is shown holding an elephant and a turtle in his claws. Both were borne aloft by Garuda and were smashed down. When Garuda was just about to eat them, they changed back to their original form as two brothers before they had been cursed by the gods and turned into an elephant and a turtle. Both conveyed their gratitude to Garuda who had released them from the curse.

Garuda was chosen as a the symbol of the Republic of Indonesia for his successful effort to free his mother from the suffering of slavery. To the Indonesian people, the mother is Pertiwi (meaning motherland), who had been set free by Garuda, her mighty son (the young generation) from slavery and subjugation. Be brave and mighty like the eagle who respected his father always (Sang Kasyapa). When Indonesians look at the sky, we see our father, Father sky.

References

Department of Education and Culture, The Project for the Conservation of Cultural Media, 'Sukuh Temple in Central Java'.

Hoop, A.N.J. Th. à Th. van der (1949), *Indonesische Siermotiven. Ragam-ragam Perhiasan Indonesia* [*Indonesian Ornamental Design*], Bandung Koninklijk Bataviaasch Genootschaap van Kunsten en Wetenschappen.

Juynboll, H.H. (1906), *Adiparwa. Oudjavaansch Prozageschrift*, 's-Gravenhage, Martinus Nijhoff.

Pusat Penelitian Arkeologi Nasional Proyek Pengembangan Media Kebudayaan (1994), 'Tangkai Cermin Dengan Cerita Garudeya Dari Masa Jawa Kuna', *Harian Buana Minggu*, 7 August.

PART 3
COMPARATIVE BATIK

11 *Parang Rusak* Design in European Art

MARIA WRONSKA-FRIEND

European interest in Javanese batik belongs to one of the most interesting, although until know little-known, aspects of the centuries-long history of European and Asian artistic connections. The encounter of European artists with this undoubtedly most famous group of Javanese textiles began almost exactly a century ago. In 1892, a young generation of Dutch artists searching for a new source of inspiration undertook experiments with introducing the Javanese technique of resist dyeing to European interior decoration, fashion and craft. A decade later, the technique became well known in other European countries, gaining significant popularity among artists and craftsmen especially in Germany, France, Great Britain, Austria, Poland and Belgium. Textiles and other objects decorated with the technique of batik featured prominently during the two European art movements, commonly known as art nouveau and art deco (1890–1930).

Together with the technology of wax-resist dyeing, the principle of Javanese aesthetics and batik designs were also introduced to European art. Numerous batiks and other decorative objects produce by European artists at the turn of the nineteenth/twentieth centuries have been decorated with designs reminiscent of *sawat, timpal, lung-lungan* and *parang rusak*, frequently executed in the central Javanese blue and brown range of colours (Wronska-Friend, 1994).

In this chapter I will analyze the influence which *parang rusak*, probably the most prominent of batik designs, exercised on European art at the turn of the nineteenth/twentieth centuries. The dynamic, wavy bands of this design became visual inspiration for a significant number of European artists. Although several Western art historians noticed the unusual sinuous motifs which appeared in diagonal arrangements in the works of some European artists at the time, in all cases they attributed them to the general stylistic tendencies of art nouveau and art deco and, until now, no consideration has been given to possible Javanese associations.

109

My research indicates that in a number of cases these dynamic, linear designs in European ornamentation were created as a result of the encounter with and fascination for the *parang rusak* design developed by European artists. *Parang rusak*, either in its totality of in brief 'quotation-like' fragments, appears in the works of such outstanding artists as Henri van de Velde, Charles Rennie Mackintosh, Henri Matisse, Paul Poiret and Koloman Moser. A large group of lesser known Dutch, German and Austrian artists also created works inspired by this design.

Designs from the *parang* group are certainly the most well known and characteristic ornamental motifs which appear in Javanese textiles. There are many variations of this design, the most famous of which are those of the *parang rusak* subgroup. They are composed of continuous, wavy bands, which are placed in diagonal, parallel rows, rhythmically filling the whole surface of the cloth. The dynamic, graphic characteristics of this design are further enhanced by the choice of contrasting colours: the rows of long, serpentine bands are white or cream and are placed on a dark background. The bands are divided from each other by a chain of small rhomboids, known as *mlinjon*.

There are several varieties of the *parang rusak* design, destined for persons of suitable social rank or worn on particular social or ceremonial occasions. *Parang* is the only batik design which stresses the gender distinction of its wearer. The smallest variant of the design, known as *parang rusak klitik –* 'small broken dagger' – is destined for ladies from noble families, while men usually wear the larger version of the same design. In its largest version, known as *parang rusak barong –* 'lion broken dagger' – the design is reserved solely for the use of the ruling sultan.

The origins of this remarkable design are obscure. In literature on the subject one can find several contradictory explanations. Some of them trace the *parang* design to the double spiral of the Dong-son culture, while others see its origins in the stylized leaves of the lotus plant (Hoop, 1949). Similar compositions of linear diagonal arrangements have for centuries been produced in India. In Indonesia they can be seen on the garments of stone statues representing various Hinduist deities in the Dieng Plateau temples and can be traced back to the ninth and tenth centuries (Sewan Susanto, 1974, pp. 295–7).

But even if the prototype of Javanese diagonal designs originated in India, on Java the designs underwent significant development and transformation, becoming a unique visual statement, unmistakably associated with the culture of the central Javanese courts. According to Javanese tradition, the creator of *parang rusak* was Sultan Agung, the powerful ruler of the Mataram dynasty at the beginning of the seventeenth century (Veldhuisen-Djajasoebrata, 1984, p. 150).

At the end of the eighteenth century, the courts of Yogyakarta and Surakarta issued several edicts creating a special group of *larangan* (forbidden, prescribed) designs which were restricted to use by the sultans and their families. *Parang rusak* features in all of these edicts: for instance in Yogyakarta, only the ruler, the crown prince and their consorts had the privilege of using cloths decorated with this design. Fabrics with *parang rusak* composed part of ritual offerings given to the spirits of the sultan's ancestors (Veldhuisen-Djajasoebrata, 1980, p. 10).

Why have the Javanese people given the greatest status to this particular design? Unlike the rest of the *larangan* (forbidden) motifs, *parang rusak* does not have any cosmological connotations; its abstract, dramatic lines have no referents to the Javanese micro- or macrocosms. The answer to the significance of this design may therefore be in its visual properties: even from a distance, the bold slanting lines of the design provide immediate recognition of the person who wears the cloth (Adams, 1970, p. 25).

It is possible that the same visual qualities of the *parang rusak* design appealed to European artists and designers active at the end of the nineteenth and beginning of the twentieth centuries.

One of the first artists who was undoubtedly inspired by the abstract dynamics of this design was Henri Van de Velde (1863–1957), probably the most famous Belgian artist from the turn of the nineteenth century who, after initially pursuing neo-impressionist painting, around 1890 shifted his interests to applied arts and architectural design. Due to his interest in the revival of craft and new decorative techniques, Belgium, after Holland, became the second European country where experiments with the Javanese method of resist dyeing had already been undertaken in the last decade of the nineteenth century.

Although Van de Velde probably made very few, if any, batiks himself, in 1896 he was instrumental in encouraging his friend and visitor to his house at Uccle, the Dutch artist Thorn Prikker, to initiate production of batik textiles. As the correspondence of Thorn Prikker reveals, both artists had an excellent knowledge of the Javanese dyeing technology and batik textiles in general (Joosten, 1972). Soon afterwards, Van de Velde designed a number of objects decorated with strong, winding lines in continuous arrangements, which bear a strong similarity to *parang rusak*. One of them is a gown designed for his wife, the hem of which was decorated with a flowing double spiral motif, closely resembling the rolling bands of the Javanese design.

Although this similarity may seem to be coincidental, resulting from the common linearity of the art nouveau style and the Javanese design, a more definite indication of the influence of *parang rusak barong* is visible in the bold, dynamic lines of the arabesque which dominates the famous poster 'Tropon' created by Van de Velde in 1898. Then, in 1900, the similarity becomes even more obvious in the series of continuous, wavy designs for woven fabrics which Van de Velde created for the Temporary Exhibition of Ladies' Garments, organized by the Kaiser Wilhelm Museum in Krefeld, Germany. The fabrics were produced in 1901 by the factory Deuss & Oetker in Krefeld (Gronwoldt, 1980, p. 242).

During research at the Van de Velde Archives in Brussels, I was able to discover direct evidence which proves that Van de Velde certainly knew *parang rusak* and that he considered using original Javanese batiks at least in one of his projects. This evidence is provided by a photograph of an original Javanese batik with the *parang rusak* design which, according to the accompanying information, was used around 1905 to decorate the interior of a house designed by Van de Velde (Van de Velde Archives 1039bis, Brussels). Probably this particular photograph led to the erroneous presumption that Van de Velde personally designed this Javanese batik, as stated by Masini in the book outlining the history of art nouveau (Masini, 1984, il. 323).

Henri Van de Velde was not only a practising artist, but was also interested in creating a theoretical framework for the new artistic movement, which later became known under its French name as 'art nouveau'. In his treatise on surface ornament published in 1902, he stressed the significance of abstract, dynamic lines and stated that 'a line is a force, filled with the energy of him who drew it' (Schmutzler, 1978, p. 212). *Parang rusak* perfectly met these requirements: with the interplay of curved lines and empty spaces, it was a well-suited design for the two-dimensional style of abstract ornamentation developed by Van de Velde and favoured in art nouveau ornamentation in general.

A similar series of textile designs, undoubtedly inspired by *parang rusak*, was produced by another Belgian art nouveau artist, George Lemmen. Around 1900, Lemmen was commissioned to produce a series of textiles decorated with the batik technique for 'La Maison Moderne' in Paris. Until now, I have been unable to locate any of these batiks in museums' collections, but two of his fabric designs are known from photographs published in the periodical *Decorative Kunst* in 1899. There is no mention of the technique in which these textiles were executed, but the designs bear close similarity to the Van de Velde *parang*-inspired fabrics. Further research is necessary to determine whether Van de Velde and Lemmen cooperated in designing textiles and interiors at that time, or whether the inspiration of *parang rusak* aesthetics had appeared in their works independently.

It is difficult to decide with the same certainly whether *parang rusak* had any impact on the works of Jan Toorop (1859–1928), probably the most significant Dutch artist active at the turn of the nineteenth century. Toorop was born in Mojokerto on Java where he spent the first 11 years of his life. This fact had a pronounced effect on his artwork. Almost all human representations created by this artist resemble the figures of the *wayang kulit* marionettes: they have a flat, frontally represented body with head shown in profile. Their stiff, thin arms are bent at sharp angles. This type of linear, graphic representation is particularly frequent in Toorop's drawings and paintings from the early 1890s, such as 'Three Brides' (1892) or 'Song of the Times' (1893). The strong sinuous lines which evolve from the hair of female silhouettes and dominate the whole composition, may again be attributed to the highly transformed stylization of the *parang* design, especially as it appears in the context of other Javanese influences. On the other hand, these overpowering continuous lines could be attributed to the general linearity of the art nouveau style.

A closer resemblance to the *parang rusak* design is visible in the poster 'Delftsche Slaolie' which Toorop created in 1898. It is also worth noticing that other designs on this poster might have been borrowed from the visual vocabulary of Javanese batik: the dress of the right *wayang*-like figure is covered with a floral geometricized motif known as *ceplok*, while in the upper left corner a large square block of abstract ornament resembles an imprint of a cap block. Strong, twisting lines of contrasting colours also appear on several book covers, designed by Toorop during the same years (for examples, see Braches, 1973). Even if *parang rusak* provided the initial impetus for this type of highly abstract, linear ornamentation, then the influence has been deeply assimilated and transformed to conform with the individual style of this artist.

Another artist inspired by Javanese aesthetics at the turn of the century was the Austrian Koloman Moser (1868–1916). A photograph of his studio interior from 1901 reveals a running frieze composed of an abstract, sinuous, lightly coloured band. In spite of the remarkable similarity to *parang rusak*, the design could again be attributed to the general decorative style of art nouveau. However, in the context of other decorative designs created by Moser at that time, which reveal definite borrowings from Javanese aesthetics (for instance, the frequent use of the *sawat* design), the probability of the *parang* influence is almost certain. Moser designed, as well, several other textiles in which lightly coloured, winding bands had been placed on a darker background (in one of the cases the band had been composed of closely placed figures of trout). These compositions may further point to *parang rusak* connections (for examples, see Gronwoldt, 1980, p. 174).

Parang rusak inspiration may also be present in the textile designs of Robert Oerley, another Austrian artist from the turn of the century. His design for woven silk and cotton cloth from 1899 entitled 'Cosmic Mist' reveals great similarity to the principles of the *parang rusak barong* composition (ibid., p. 182).

A decade later in France, Paul Poiret (1879–1944), the most famous avant-garde couturier in Paris at that time, designed a series of evening gowns and coats, executed in the batik technique. The garments are relatively well documented in numerous drawings and photographs publishing in contemporary French fashion and art periodicals, as well as in sketches by George Lepape, a designer who collaborated with Poiret at that time.

Of particular interest are two evening coats, both executed in the blue and brown central Javanese colours and decorated with Javanese designs: one of them with a row of *tumpals*, while the other, entitled 'Battick', with wide, slanting bands filled with alternating rows of rhythmic lines and circles. It is probable that the design represents a simplified version of *parang rusak*, in which the continuous serpentine line has been transformed into multiple repetitions of parallel stripes. The general principle of composition, however, remains the same as a Javanese batik: the oblique rows of the abstract, linear design dominate the whole surface of the garment, filling it with powerful dynamism.

Parang rusak consistently appears in the textile designs and on the paintings of one of the most prominent British artists, Charles Rennie Mackintosh (1868–1928). A famous architect, painter and designer, Mackintosh as an active member of the artistic group that emerged in Glasgow at the turn of the nineteenth century.

In 1915, in the latter part of his life, Mackintosh moved to London where, during the following eight years, he designed several series of fabrics for two textile firms, Foxton and Selfton (Billcliffe, 1993). A group of abstract diagonal designs in vivid colours feature prominently in Mackintosh's textile compositions. In many of them one can clearly recognize the characteristic rhythm and continuous flow of *parang rusak*. It is known that Mackintosh used to collect Javanese batiks (Robinson, 1969, p. 14) and undoubtedly, his collection also contained several designs from the *parang* group.

Parang rusak appears only sometimes in Mackintosh's textiles in a version close to its Javanese prototype. More frequently he translated the characteristic bands of the design into rows of tulips, daisies, stylized leaves, peacocks' eyes or even human figures (for example, the 'Odalisque' design). In the other type of stylization, Mackintosh proceeded towards creating purely abstract, simplified zig-zag figures, which were given names such as 'wave patterns' or 'wave and zig-zag' designs (for examples, see Billcliffe, 1993).

During the same period Mackintosh painted a series of watercolours which represented various floral arrangements. Frequently, in the background of these paintings, the artist repeated his textile designs, usually wavy bands inspired by several types of *parang rusak* design. The most well known of these are 'Anemones' (1916), 'Peonies' (c. 1919), 'White Roses' (1920) and 'Cyclamen' (1919–20).

Unlike Van de Velde and Lemmen, who favoured *parang rusak barong* and stylized the design, stressing its soft and flowing characteristics, Mackintosh preferred the sharp lines of *parang curigo* and emphasized the bold, rhythmic, geometrical qualities of this design. This type of stylization created already in 1915–16 made him one of the pioneers of the art deco style.

Finally, the last of the artists to be presented, and certainly the most famous one in whose work one can trace the influence of *parang rusak*, is Henri Matisse (1869–1954). Throughout the many years of his creative life, the style of his paintings underwent significant changes, shifting from realistic representations towards bold, abstract designs. In his works created between 1908 and the late 1920s, bright colours and powerful background patterns dominate the compositions, combining people, objects and designs into one ornamental block. Many of the strong designs overriding these pictures had been borrowed from Indian and Islamic textiles, of which Matisse had a significant collection. Matisse was fascinated for many years with oriental art and was supposed to have admitted that 'revelation always came to him from the East' (Gowing, 1979, p. 111).

Although the *parang rusak* design would certainly comply with the strong, overpowering patterns of Matisse's paintings, until now I have been unable to find definite evidence of *parang rusak* or other Javanese designs appearing in any of his work created at this time. Although in 1906 in one of his woodcuts entitle 'Nu, le grand bois' we find a bold zig-zag design which could be a transformation of the *parang* design, the similarities are too distant to be used as evidence. In 1929, in one of his etchings 'Figure seated in interior', there appears a similar serpentine motif which might or might not be attributed to the *parang* design.

However, unmistakable proof of Javanese connections appears in a series of Matisse drawings published in 1936 under the title 'Cahiers d'Art'. The most frequent subject of these drawings is a nude female figure, placed in an interior filled with decorative textiles, flowers, pottery and birds. Matisse displays in these works a mastery of strong, pure lines, which combine all elements of the drawing into an integrated, linear representation. The excitement resulting from the discovery of the well-defined, graphic lines of *parang rusak barong* has been expressed in several drawings, where realistically represented outlines of the design appear as a canvas for displaying the figures of the models.

At the turn of the nineteenth century a number of Javanese designs had been introduced to European art as a result of experiments conducted with the batik technique by European artists, although *parang rusak* is the only design which has transgressed the medium of wax-resist dying. In consequence, we find this design applied to woven and printed textiles, in interior decoration, in paintings and in drawings of several of the most prominent European artists of the time. How can one explain the significant international career of this particular design? Certainly it was not the ritual and ceremonial significance of the design in Javanese culture which drew the attention of European artists, as few, if any of them, were aware of the special importance given to the *parang* designs on Java. The fascination results probably from the formal qualities of the design: its abstract, sinuous lines which paralleled the decorative style of art nouveau, while its boldness, rhythm and colour contrasts appealed to the next generation of artists, working in the stylistic convention of art deco.

The aesthetic qualities of the *parang rusak* design thus not only achieve the highest acknowledgement of the Javanese people, who restricted its use to the most respected and significant members of their society, but also excited and inspired a number of significant European artists at the turn of the nineteenth century. The history of this noble design confirms the existence of aesthetic phenomena which are universally accepted by people across times and cultures.

References

Adams, M. (1970), 'Symbolic Scenes in Javanese Batik', *Textile Museum Journal*, 3 (1), pp. 25–40.

Billcliffe, R. (1993), *Charles Rennie Mackintosh. Textile Designs*, San Francisco, Pomegranate Books.

Braches, E. (1973), *Het boek als Nieuwe Kunst*, Utrecht.

Gowing, L. (1979), *Matisse*, London, Thames and Hudson.

Gronwoldt, R. (1980), *Art Nouveau. Textil Dekor im 1900*, Stuttgart.

Hoop, A.N.J. Th. à Th. van der (1949), *Indonesische Siermotiven. Ragam-ragam Perhiasan Indonesia* [*Indonesian Ornamental Design*], Bandung Koninklijk Bataviaasch Genootschaap van Kunsten en Wetenschappen.

Joosten, J. (1980), *De Brieven van Johan Thorn Prikker aan Henri Borel en anderen 1892–1904*, Uitgeverij, Heuff Niuewkoop.

Masini, L.-V. (1984), *Art Nouveau*, London, Thames and Hudson.

Robinson, S. (1969), *A History of Dyed Textiles*, London, Studio Vista.

Schmutzler, R. (1977), *Art Nouveau*, London, Thames and Hudson.

Susanto, S. (1974), *Seni Kerajinan Batik Indonesia*, Jakarta.

Veldhuisen-Djajasoebrata, A. (1980), 'On the origins and nature of larangan: forbidden batik patterns from Central Javanese principalities' in M. Gittinger (ed.), *Indonesian Textiles. Irene Emery Roundtable on Museum Textiles 1979 Proceedings*, Washington DC.

Veldhuisen-Djajasoebrata, A. (1984), *Bloemen van het Heelal*, Amsterdam, A.W. Sijhoff.

Wronska-Friend, M. (1994), 'Batik and European textiles at the turn of the 19–20th c.', paper presented at the conference *Indonesian and Other Asian Textiles: A Common Heritage*, Jakarta, National Museum.

12 Javanese and Indonesian Batik in Germany

IRENE ROMEO

During the years 1970–90, batik experienced its high point in Germany and Europe as a whole. It was widely taught as a hobby in adult education and arts and crafts schools.

Professional batik artists began to establish themselves and serious batik books for amateurs and scholars were published. These included, among others, Annegret Haake (1984), Hildegard Santner (1979), Miep Speé (1977), Ursin and Kilchmann (1979) and Yding (1979). Haake's contribution was well researched and included an overview of the development and history of Indonesian batik; she commented 'Occidental batiks want to be considered as artistic expressions ...'. An effort was made to incorporate batik techniques and it was taught as part of the official art curriculum in schools. Both in hobby arts and crafts programmes and in the school curriculum, one did not only teach *pelangi* (tie dye), but emphasized instead the use of wax-resist.

The historical development of batik was acknowledged and this led directly to Java. Exhibitions brought both teachers and students into the museums. The catalogues of these exhibitions indicated the essential directions. Authors elucidated the historic batik practice in Java. They opened up the cultural and historical background of Javanese batik to teachers, students and professional batik artists and for the buyers.

The studio of A. Wolff and I. Romeo taught batik for over 10 years to teachers and social workers. Annually about 60 participants in seminars learned the basic batik. Wolff and Romeo were appointed to teach batik at the University of Bielefeld.

In museums and exhibitions both old and new Javanese batiks were shown. These presentations usually generated great public interest. European and German batik artists were exhibiting in private exhibitions which sometimes also included examples of old and new Javanese batiks. In this context a particular phenomenon could be observed: the public reacted very positively to scholarly slide presentations on Javanese batiks, their origins and history.

European and German batik artists realized that their work sold best after a joint exhibition of Javanese and modern batiks and a lecture. The simultaneous availability of Javanese batiks did nor affect the sales of the German batiks. One should note that Indonesian batik always ought to be an integral part of a combined exhibition. During the batik boom in Germany, Indonesian batiks sold very well in exhibitions which featured them exclusively. Of course this was affected by the prices. If prices were above the usual limits, the items for sale were only hesitantly bought. Because in these years between 1970 and 1990, courses in batik were offered in most secondary schools and art clubs, batik was so 'in' that almost all events and exhibitions which featured batik were well attended. Batik as a hobby reached its high point at this time.

The public learned very quickly to evaluate the batiks and to take advantage of access to the historical batik motifs from Indonesia. They transferred a certain set of quality expectations to German/European batik work. Many batik professionals could accept that and align their work to public expectations. In many schools batik came to be a professional specialization, a major. The further education of teachers and students in this major contributed substantially to the development of artistic quality standards in batik.

In the late 1990s this situation has substantially changed in Germany. Today there are hardly any exhibitions of batik. Museums, schools and universities and other publicly funded institutions have to struggle with budget cuts. In the adult education, hardly any batik courses are offered. In teacher preparation programmes the study of batik has been eliminated.

It was practically a miracle that in 1996 I could organize in the museum of Nienburg, a county seat in the state of Lower Saxony, a batik exhibition of old and new batiks from Indonesia and work by a professional German batik artist. The exhibition was on for three weeks and had a high rate of visitors. The Indonesian batiks were not for sale, though the German ones were, but none were sold. The present economic situation of Germany has resulted in lowering purchasing power and therefore art and craft are experiencing a state of stagnation.

This, however, is not the only reason why batik has little chance of success in late twentieth century Germany; it has been replaced by silk painting. Society in the late 1990s is characterized by 'joy and fun', but batik is very labour intensive and cannot satisfy this demand even though 'joy and fun' are not absolutely excluded from batik work.

We can therefore summarize that economic and social processes are presently impacting negatively on batik in Germany, affecting Indonesian as well as German batik.

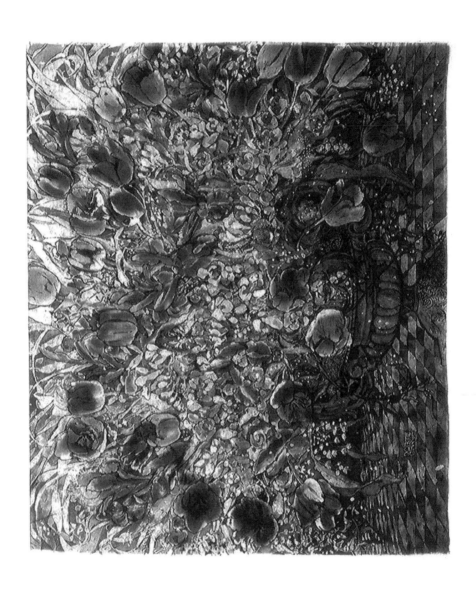

Illustration 12.1   Dorte Christjansen: 'Spring Bouquet' – batik on silk, 23 x 32 in.

Illustration 12.2 Dorte Christjansen: 'Bordertown' – batik on silk, 32 x 23 in.

**Illustration 12.3   Dorte Christjansen: 'Owens Valley' – batik on silk,
17 x 12 in.**

Batik will be rediscovered. When schools, teacher training programmes, secondary schools and museums are able to move more freely, especially financially, batik will find its significance again. Perhaps with new impulses. The same would be true for merchandising methods.

For years Rudolph Smend has initiated exhibitions and shown old and new Indonesian/Javanese batiks in his gallery in Cologne. He also showed many modern representatives of batik from around the world. Perhaps he may be able to find and embark upon new ways.

Effects of German Tourism in Indonesia

Tourism may be able to contribute to the revitalization of interest in traditional batik and vice versa: traditional batiks may contribute to the revitalization of batik-tourism. Batik, however, cannot compete with sensational travel opportunities, like for example the spectacular hiking tours of the mountain Gunung Rinjani on the island of Lombok or hiking tours on other volcanic mountains or diving in the coral reefs along the north coast of Sulawesi. But the exceptional subtleties of Indonesian batik may be able to find a very different clientele.

It is important to recognize the following: there are two aspects of the understanding of traditional Javanese batik. If you are involved with Indonesian/Javanese batik you will notice that the lines of wax have two functions. First, the lines that are applied with the *canting* separate and enclose clearly the intended forms. Second, the ornamental motifs also have formative and creative potential. Lines that are characteristic for all sections may indicate schematic abbreviations; they let the essential be recognized, but exclude variable individuality. Batik reflects the formative phases of the old Indonesian tradition, of the religious high cultures and the court cultures, and the communication system created by the traditional cultures. This understanding of cultural context and history should be taught and, in the context of exhibitions, explained. Usually the public encounters traditional Indonesian/ Javanese batik in schools, museums, special seminars and events and wants to know more. To address this need, the batik studio of Wolff-Romeo took the opportunity to invite their students to travel to Indonesia. In the 1970s and 1980s from the relatively small circle of this studio, 12 tours, each with 10–15 participants, were conducted. It should be possible to organize such projects with different organizers and find sufficient interested participants.

Production of Batik Accessories in Germany

Production, exhibition and sales of batik scarves has a tradition in Germany. They were the main item produced by the hobby batik movement and therefore in adult education and various arts and crafts courses. This carried over into the production of shawls, borders and ribbons for clothing and bags. In the batik boom years these types of accessories were especially successful. The participants in the courses not only achieved with these items a style of clothing, but more than that, contributed to a sense of well-being and self-validation on the part of the wearer. When such clothes were shown in exhibitions they sold well at good prices even if they were above average in quality.

Batik Fashion in German-speaking Areas

Most of the professional schools seek innovative ideas in the curricular structure of fashion and textile design. Experimental processes are usually supported and recognized and deal with two- as well as three-dimensional design problems. Particularly important in this context are the materials, colours and designs for special events or for everyday clothing prepared for ready-to-wear production. Certainly international competitions are influential. Fashion is created with consideration for particular qualities: affirmative, creative, high-class, eye-catching, impressive, spectacular, special and original. Why could batik, even traditional Javanese batik, not be used with some vision? Silk has found acceptance again. All these are possible starting points of which one should take advantage. To whom is this challenge addressed? All of you with contacts to designers, fashion manufacturers and whoever is active in the area of fashion. Good ideas and flexibility are needed.

Of course there needs to be purchasing power and the desire to buy must be encouraged, particularly in the ways mentioned.

In 1995 we invited Iwan Tirta to exhibit his wall-hangings in the Museum Nienburg; at the same time batik pictures of a German batik artist were to be shown. Iwan Tirta had agreed. The entire exhibition had been planned. Unfortunately, the state of Lower Saxony cut the funds for the museum for the year 1995. When these were released for the subsequent year we no longer were able to get the show because of Iwan Tirta's exhibition schedule.

Optimism should win out and new plans need to be drawn up. Creativity is necessary in our organization and all of us can contribute to this. Perhaps this chapter can help to inspire the ideas, discoveries and results.

References

Billeter, E. (1970), *Aussereuropäische Textilien Sammlungskatalog 2*, Kunstgewerbemuseum, Zürich (o.J.).

Feldbauer, S. (1988), *Bathik – Simboli magici e tradizione femminile a Giava*, Milan, Electa Spa.

Haake, A. (1984), *Javanische Batik. Methode, Symbolik, Geschichte*, Hanover, Schaper.

Irwin, J. and Murphy, V. (1969), *Batiks*, London, Victoria and Albert Museum.

Jasper, J.E. and Pirngadie, M. (1916), *De Batikkunst. De Inlandsche Kunst-Nijverheid in Nederlandsh-Indie, Vol III*, S'Gravenhage, Mouton.

Kahn, Majlis, B. (1984), *Indonesische Textilen, Wege zu Göttern und Ahnen*, Deutsches Textilmuseum Krefeld, Rautenstrauch-Joest-Museum für Völkerkunde, Köln, Wienand.

Lee Chor, L. (1991), *Batik Creating an Identity*, Singapore, National Museum.

Loeber, J.H. (1926), *Das Batiken*, Oldenburg, Gerhard Staling Verlag.

Maxwell, R. (1990), *Textiles of Southeast Asia – Tradition, Trade and Transformation*, Australian National Gallery.

Mylius, N. (1964), *Indonesische Textilkunst – Batik Ikat und Plangi*, Vienna.

Nabholz-Kartaschoff, M.-L. (1970/71), *Batik*, Basle, Museum für Völkerkunde und Schweizerisches Museum für Volkskunde.

Oss van, F. (1996), *Batik. The soul of Java*, Textilmuseum Tilburg/Nederland and Jakarta/Indonesien.

Raffles, T.S.B. (1978 [1817]), *The History of Java*, Kuala Lumpur, Oxford University Press.

Romeo, I. (1996a), 'Patchwork-Quilts und Batikbilder', *Textilkunst*, 23, 4, pp. 179–81.

Romeo, I (1996b), 'Batiken der besonderen Art', *Textilkunst*, 24, 1, pp. 39–40.

Romeo, I. (1996c), 'Batik – die Seele Javas', *Textilkunst*, 24, 4, pp. 203–4.

Romeo, I. (1996d), 'Batiken aus Baumwolle und Seide – Geschichte, Auffassung und Phänomen der javanischen Batik', *Katalog*, Museum Nienburg/Weser, pp. 63–102.

Romeo, I. (1997), 'Seide und Batik', *Textilkunst*, 25, 1, pp. 47–8.

Rouffaer, G.F. and Juynboll, H.H. (1914), *De batikkunst in Nederlansch-Indië en Haar Geschiedenis*, Utrecht.

Sandtner, H. (1979), *Schöpferische Textilarbeit*, 4th edn, Donauwörth, Auer.

Speé, M. (1977), *Traditionele en Moderne Batik*, Canteleer bv, de Bilt.

Tirtaamidjaja, N. (1966), *Batik, Pola & Tjorak (Pattern & Motif)*, Jakarta.

Ursin, A. and Kilchemann, K. (1979), *Batik*, Stuttgart, Haupt.

Veldhuisen, H.C. (1993), *Batik Belanda 1840–1940 (Dutch Influence in Batik from Java History and Stories)*, Jakarta.

Veldhuisen-Djajasoebrata, A. (1972), *Batik op Java*, Rotterdam, Museum voor Land-en Volkenkunde.

Völger, G. and Verruca, K. (1985), 'Indonesian textiles', symposium, Cologne, Rautenstrauch-Joest-Museum.

Yding, K. (1979), *Batik*, Borgen.

13 Reflections on Moscow Batik

JOACHIM BLANK

Before interpreting contemporary Russian batik art through the work of its most outstanding artists, it is useful first to begin with a short outline of Russian painting from its beginnings to the present day.

In the eighth century painting was introduced to Russia from Byzantium. By the eleventh century, a northern-Byzantine style with its own plant, leaf and animal design had developed out of these influences. At the same time in the eleventh century, Persian motifs contained in early Islamic art came in from Mongolia. This had a significant impact upon Russian painting, particularly in Uzbekistan, Turkmenistan and Azerbaijan. In the fifteenth century, the floral ornamental style disappeared suddenly and a more geometric ornamental trend emerged, influenced by early Christian paintings.

Starting in the twelfth century, icon painting, with its origins in northern Russia, became increasingly influential and was to dominate Russian art into the fifteenth century. This epoch of Russia's religious art was rooted in the Novgorod School, the centre of the Russian Orthodox style of painting. There was no secular painting in Russia until the arrival of Danish, Swedish, Italian and French painters, who travelled across Russia in the seventeenth century. Russian genre painting achieved its first high point in the early eighteenth century under Czar Peter the Great. Landscapes occupied a central position within Russian painting at that time. A deep sense of mysticism and poetic symbolism characterizes Russian painting to this day, even leaving its imprint on the style of the nascent Russian avant garde at the turn of this century. The epoch of modern art in Russia was dominated by Suprematism, its main proponent being Malevich. With Leninism, and later Stalinism, the arts in Russia suffered a levelling process (*Gleichschaltung*) which left a deep impression and drove many of the avant garde artists abroad, including Kandinsky, Chagall and Yavlenski.[1]

It was only in the 1960s that a more individual, freer expression could develop within the visual arts. With the advent of *perestroika*, Russian painting re-emerged without an ideological superstructure.

The Moscow batik scene can be divided into two fundamental approaches.

On the one hand there are those artists who base themselves consciously on different styles and epochs in painting, using the trendsetting artists as an orientation point for their own work. On the other hand, there are artists who, by the way they encounter and reflect nature, may be linked to the great tradition of Russian landscape painting, typified in the works of Levita, Vasilyev and Shishkin.

However, the influence of past masters has not had the same obvious impact on the work of this latter group of artists. Rather, they are in search of their own artistic identity without revealing any link to a recognizable style or school of painting.

Nadia Golubtsova and Tatiana Shichireva, both Moscow batik artists, belong to those who model themselves on past traditions. The influence of Russia's Suprematist painters of the 1920s is clearly discernible in the large format batiks of Golubtsova. Her style is consciously inspired by Malevich, whose geometric-constructivist compositions were dominated by seemingly weightless, sailing shapes that create visual depths. These are also typical for Golubtsova who merges this influence with early cubist elements. Golubtsova subordinates her extensive application of colour to enhance the play of shapes in her work. Colour contrasts, too, are determined by the existing geometric composition.

Nuances within Tatyana Shichireva's cheerful fairytale style suggest the lively worlds depicted in fourteenth century Gothic painting, known in Western art history through the *Livres d'heures* or the twelfth century early Gothic Zurich *Manesser Handschrift* (Manesser manuscript). Shichireva's masterly filigree batik technique enables her to create images which tell of this artist's fascination for the world of the troubadour, the street artists, cavaliers and knights of the Gothic age. Shichireva's lively coloration and the poetic composition of her work actually invite the observer to become immersed. The influence of Gothic painting in Shichireva's batik becomes apparent in another aspect of early Gothic painting, which is also consciously integrated into this artist's work; the lack of exact perspective in the composition of the picture.

Sergei Davidov, a batik artist born in Novorossisk, refers in his work to ancient Russia's icon and religious fresco paintings. His pictorial language points to that era's metaphors laden with deep mysticism and powerful symbolism. This, however, is counterbalanced in a significant way by the inclusion of landscapes and people. Davidov's historic visual models do not dominate his batiks but rather are reinterpreted and play a new role within his compositions. In this way, completely new pictorial realities are created.

**Illustration 13.1   Tatyana Shichireva: 'Troubadour' , 75 x 80 cm,
1996**

**Illustration 13.2   Sergei Davidov: 'Uglitali' – batik on cotton,
60 x 60 cm, 1995**

Davidov's imagery contains encoded messages and interpretations of subjects which are neither harmless nor superficial. The artist's central pictorial meanings are conveyed through the occasionally threatening interpretation of the depiction of people. These stem from his immediate and subjective experience as an artist in 1990s Russia. Davidov's compositions are reminiscent of collages and thereby develop a special appeal, combining different picture sequences into a whole.

The impact of figurative surrealism is unmistakable in the batik work of George Boulytchev. In his art, the world of rational experience is expanded into pictorial space in which perceived reality is transformed into the unreal. Alienated architectural forms break out of an almost monochrome visual background and lead the observer into dreamlike pictorial worlds. Sparing use of colour lets the condition of unreality appear more sharply. Boulytchev's batik style also shows that he discovered textile art through tapestry weaving. This in turn makes his batiks appear two-dimensional, without diminishing their appeal.

Among the batik artists who follow the Russian landscape tradition Yuri Salman and Alexander Talaev are the principal figures. The flowing backgrounds of Salman's batik landscapes are an impressive interpretation of the reflecting surfaces of Russia's endless seas and horizon-wide lakes. Pushing into these works like a backdrop are shadows of the mighty antipodes of the Russian countryside – mountainscapes that recall the Urals and the Caucasus. A fascinating feature in Salman's work is his peculiar and powerful illumination, bringing to mind strong reflections of sunlight at dusk and the pale yellow light in the steppe. The interplay of Salman's partly geometric compositions with his powerful colour palette exude meditative quiet which involuntarily attract the observer.

Alexander Talaev's large format batiks, on the other hand, cannot be easily identified as interpretations of landscape at the first glance. They are more inclined to bring to mind the cast landscapes of the Baikal region locked in eternal ice. Talaev's flowing forms no longer allow for the topographically exact location of his models. This is furthered by a specially developed batik marbling technique used in combination with a special wax pouring technique. Here, modern art has fixed the border line to abstract painting; Talaev's batik work is located where landscape can no longer be defined by concrete reality. Talaev's work has ensured him a place among the most outstanding of Moscow's textile artists as his landscape interpretations are unmistakable and not too academic.

Alexander Damaskin holds a special place among those Moscow batik

Illustration 13.3   Yuri Salman: 'Landscape' – batik on crêpe-de-chine, 1995

artists who belong to the landscape school. The characteristic style of his work, the graphic-like technique and the composition of his images, all combine to create very strong associations with Tun Yuan's landscape painting. The Chinese artist Tun Yuan lived and worked in Nanking during the second half of the tenth century. Damaskin, like Yuan, has discovered in his landscape batiks an interpretative form which gives a sense of quiet and contemplation without sweetness in the usual sense. Damaskin does not idealize his landscapes but removes their starkness by covering the mountain peaks in mists. Here too Damaskin's work displays a strong affinity with Chinese painting. In his floral ornamental work there are references to tropical plants and flowers which echo in their intense colouring of contemporary Indonesian batik. Damaskin's floral batiks display a distinctive transparent delicacy which, in their interplay with colour and shape, produce an almost fragile creation.

Victor Kosiak too cannot easily be classified as belonging to either of the approaches mentioned earlier. Kosiak's subjects are very different and place him apart from all the other Moscow batik artists. It is true to say that his style may be identified with surrealism, but the iconography of his batik takes the observer into the world of demons and chimeras, somewhere between the Himalayas and the Indian Ocean.

His visions are drawn from the body of myths and legends in the Western and Russian worlds. However, anybody familiar with Indonesia will associate corresponding images from the works of Wayang Topeng or Barong Keket when looking at Kosiak's batiks. Kosiak's highly alienated animal images are reminiscent of a world of lizards and iguanas. And those shapes which often cut through his snakeskin-like abundance remind one of elephants' tusks. Fantastic realism, defined as a surrealist offshoot by art historians, has created a breathtaking visual world in Kosiak which leaves a lasting impression on the observer. Kosiak's overflowing work, which seems obsessed with detail, creates an inexhaustible potentiality of a world that does not define itself externally or internally but finds its place exclusively in the image.

A comparison of Russian and Indonesian batik arts is not an easy task. Due to the vastly different artistic traditions and cultures it is necessary to look for possible points of comparison that will enable us to draw insights.

A useful approach is to examine myths and stories as conveyors of meaning in both artistic directions without diminishing the distinctiveness of Russian or Indonesian batik art. A study of Javanese batik art from the Indo-European period reveals examples of the narrative style depicted on the sarongs from Mrs Metzelaer and Mrs Van Oosterom's batik workshops. I mention by way of example 'Little Red Riding Hood' and 'The Battle of Lombok' or indeed

**Illustration 13.4   Alexander Damaskin: 'Landscape' – batik on silk,
70 x 80 cm, 1995**

**Illustration 13.5   Alexander Damaskin: 'Flower' – batik on cotton,
70 x 110 cm, 1996**

«East Dream»

Illustration 13.6   Alexander Kosiak: 'East Dream' – batik on cotton, 200 x 900 cm, 1995

Mrs Franquemont's 'The Theatre'. Going beyond the story-telling level in the quest for points of comparison, we discover in Tatyana Shichireva's pictorial worlds a very similar historical approach in her work.

Davidov's visual worlds also point to those poetic variations, although highly alienated and not always easily recognizable at first glance. Kosiak's batiks, on the other, hand depict pictorial worlds with their magical and demonic expressions which let one presume an affinity with the gods, mythical creatures and spirits of the Indonesian-Hindu region. This does not imply a naïve imitation but represents an integral part of Kosiak's individual style.

The highly-crafted, century old artistic tradition of Javanese batik has always been deemed a highly respected Indonesian art form. It has inspired countless European artists to practise this variant of textile art. However, in contrast to Indonesia, batik has not been able to establish itself as an independent art form in the West in the way it would have deserved in view of the noteworthy examples of Russian batik art. Art historians in the West have classified batik art very much as a craft and therefore it remains inaccessible as an independent art form to the visual artist.

Note

1   Jawlensky.

14 Batik as Both Art and Craft

RITA TREFOIS

For a lot of people in the Western world, certainly in Europe and especially in Belgium, batik is an occupation that often starts as a hobby, becomes a craft for some of them and finally a full-time job as an art for a few of them. But people in general don't always appreciate the technique of batik as it is re-evaluated and utilized as an art-medium in Belgium, although there seems to be a growing awareness.

It is well known that the Indonesian batiks inspired many Western artists. Indonesian batik is known worldwide on account of the skill and precision invested in traditional fabrics and the innovative approaches of contemporary artists.

I don't remember why or when I was attracted by batik, but as neighbours of the Dutch, we Belgians are more or less influenced by what happens there. Since my childhood I have been attracted by fabrics. I painted on cloth, helped by my drawing teacher at school and my father, a famous photographer. Later I studied textile chemistry, worked in a lab, got married and raised children. Suddenly I missed the artistic background of my childhood. The painting, drawing, textiles and chemistry all together drew me to batik.

For me there was a long experimental period based on my knowledge about fabrics and colours. I worked on my own and became obsessed by the technique: all white fabric at home was decorated in such a way that my husband dared not put a white shirt in the laundry, lest it be decorated with batik. I was very prolific and so learned a lot in that period. In the beginning, working on my own, I adopted a somewhat traditional style, partly because I was so meticulous. I was self taught because there was no education available at the time. I wanted to do everything in the right way. Having a good technique from the start was very important to me.

My work was figurative in the beginning, especially flowers, but rapidly it turned into more abstract designs. I experimented with portrait, but because my father was a photographer and my sister was drawing and painting portraits, it remains only a brief experience I left for the more abstract work.

Suddenly there was some influence from outside: I learned about Noel

Dyrenforth from England, through his book, the first good one about batik. I also heard about Rudolf Smend from Germany, because of his gallery in Cologne. Another factor was the hippy trend that made batik popular.

Without knowing Noel Dyrenforth personally at the time, he influenced a further evolution in my abstract style. To see that somebody could use this technique as an art medium was certainly a stimulation, not only for myself but for others.

Rudolf Smend was a good provider – almost the only one – of tools and equipment. Also important were his activities concerning batik: the exhibitions he organized and also the study trips to Indonesia. But the hippy trend brought a lot of inferior batiks onto the market, eroding aesthetic standards. Batik had become widely associated with hippy fashion and, following the demise of this trend, interest in batik waned. Suddenly nobody wanted batik; the good days of batik were gone.

But I was so involved with this technique, I couldn't stop experimenting. Fortunately I was not the only one. Several of us were continuing in their own places, although there is not much contact between batik people, certainly in Belgium.

My approach changed, blending colours against plain colours. I did not use solid waxing any more but wax layers with structure and transparent effects, as well as different presentations. All those experiences during all those years meant that I had a strong technical base for my artistic aspirations. Important in my own designs are colour and shape in order to obtain vigour in the complexity of composition. This evolution continues.

I started teaching batik again. I was doing it already in the 1980s and by the 1990s was really fighting to show that batik was really an art medium and not 'only' a craft as it was often seen by the general public. I teach my students good techniques and I try to give my enthusiasm for this unique art to my students. To be an excellent craftsman is necessary to become an artist. And that is why it is very important to provide good information and good training in order to obtain good quality.

Actually in our art academies, students are not really learning technique any more. Besides, batik is not appreciated enough to be included in the curriculum. It is a fact that the teachers are not always familiar with batik. That's why interested students are searching out private courses. But it is difficult to separate the wheat from the chaff and a lot of interested people become discouraged by the bad education and information about techniques and knowledge about materials and dyes.

It is important that batik is re-evaluated, and it is happening, but everything

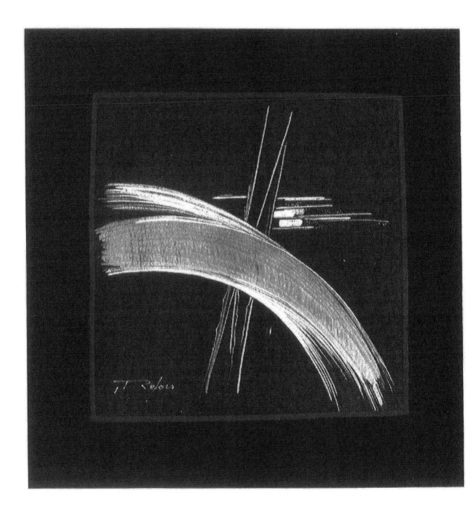

Illustration 14.1 'Rondo' (1995): wax-resist on silk, 1.4 x 1.4 m, machine patchwork, handquilting; fibre reactive dyes

Illustration 14.2 'New Outlook' (1995): wax-resist and discharge on cotton, 0.6 x 0.7 m; fibre reactive, naphtol and indigosol dyes

Illustration 14.3 **'Bursting' (1995): wax-resist and discharge on silk, 0.8 x 0.85 m; fibre reactive dyes**

Illustration 14.4 'Paris' (1997): wax-resist and discharge on silk,
0.55 x 0.75 m; fibre reactive dyes (Guizhou
Provincial Museum, Guiyang, China)

is going so quickly that I am afraid we are doing too much to bring batik to public notice. It sounds contradictory but it is a delicate matter. The danger is near that batik becomes common and loses its value as happened in the 1980s. We have to be careful and not overdo promoting batik. Batik is so rich that we have to cherish it. Promotion can be profitable for high quality batik because we have to show the 'customer' the difference, we have to educate him. Let us, please, preserve the artistic side of batik.

Because we are also talking about tourism, it is ncessary to obtain the colours from factories. With the actual environment laws in a lot of countries, those factories are not allowed any more to deliver the dyestuff to us 'private' persons.

There must be the possibility to organize markets for the professional dyers beyond the industrial circuit. I am wondering how long it will be possible for us to obtain the high quality colours we need and to use them without transgressing those laws. I think it is worthwhile to ask chemists and ecologists. We must have the possibility of using the colours in the right way, not only for the environment but also to obtain good quality in our batiks. To be a qualified dyer, it is not necessary to be a chemist, though good education and information is vital in training qualified crafts and art people.

Recently I started with a slide lecture for my students to show them what good batik looks like and to try to encourage their artistic potential. I am also preparing a book with the same purpose. I know that I am not the first one to publish a book on the subject but hope to present a refreshing or novel approach.

With the results of my activities over more than 25 years I hope I shall contribute to the appreciation of the batik technique in general and batik as an art in particular.

Reference

Dyrenforth, N. (1988), *The Technique of Batik*, London, B.T. Batsford.

15 Malay (Perak) Embroidery in Contemporary Batik

NAZLINA SHAARI

Introduction

Malaysian batik is a decorative art of tremendous beauty. However, its versatility lies in a number of factors. It is at once wearable, practical, vibrantly-coloured and a piece of art. It can take on many forms such as a simple cotton sarong, a formal lounge shirt, an elegant *baju kebaya* or even a bedspread or a painting.

The history of batik is as interesting as the material itself. It is believed to have been founded in Egypt in as early as 1500 BC although there are also claims that it originated in India. Whatever its origins, the ancient art of batik printing is most practised in Indonesia and Malaysia.

Malaysia is well known not only for batik but also for other textile crafts such as embroidery and *songket* (supplementary weft). For example, gold embroidery or *tekat*, is almost forgotten by new generations. Most of them think that this craft is just for special occasions when it is presented during wedding days or royal ceremonies. For me, looking at this craft gives me the impression that the richness of this craft should be explored, especially since batik has become a popular art. It is to show how transformation of culture exist in modern culture.

History

At first batik was applied to home-made cottons and calico but, through trade, some fine muslin reached the oriental bazaars. The finely woven quality of this cloth was perfect for batik.

The Indonesians and Malaysians also have their own unique forms of cloth design. Most of their designs are composed of stylized leaves, flowers and animals. Each set of designs is interpreted with reference to the ritual and social situations in which it is presented. A cultural study examines beliefs and values and how these are reflected in artistic endeavours. Exposure to foreign contacts has further enriched the distinct textile traditions. Striking hook-arrow and zig-zag motifs in white and red against a blue or purplish ground comprise the key elements in the magnificent traditional textiles. Gold fabrics called *kain songket* are perceived as symbols of wealth and prestige and provide a fitting display of affluence at important ceremonial events.

Third World cultures are a rich source of textile designs. Patterns taken from the raffia cloth of the Shoowa of central Africa, *pua kumbu* from Borneo, batik cloth from Indonesia and many other non-Western sources are copied and adapted into designs. Cloth is locally manufactured for export as base material for fashion apparel, utilizing time-consuming traditional techniques such as *ikat*, batik and *pelangi*, which are combined with embroidery or appliqué.

Legacy of Malaysian Textiles

Malaysia's traditional arts and crafts embrace certain styles and values from which many lessons may be learned today. Yeoh Jin Leng, guest curator, writing in the introduction to the exhibition '"Fiber, Form and Beauty", Contemporary Fiber Art of Malaysia', says that the variety of Malaysian textile displayed by the cloths of gold and silver, *cindai*, *kain pelangi*, *kain songket*, *limar*, batik, *pua kumbu* of Iban and *tubau* of Bajau bears witness to the rich heritage in textile art of this country.

Batik, Malaysia's hand-drawn fabric design, has evolved from cotton through rayon and crêpe-de-chine to Jacquard silk. It has also added a new dimension to the Malaysian fashion and identity by moving from prosaic earthstone to vibrant psychedelic hues.

Batik has undergone a quite spectacular evolution. In terms of fabrics, this unique hand-drawn design has graduated from presenting itself on cotton to rayon and crêpe-de-chine to the present range on Jacquard silk where colours concerned the usage of hues on batik has metamorphozed from dull earthtones to psychedelic ones.

What seems striking is the rapidity with which art forms and styles have yielded to various influences, notably commercial viability and the demand and the specifications of market forces. This exposure provides good opportunity for Malaysians to promote their cultural identities.

New styles and techniques deployed on all kinds of cloth have given birth to an attractive and innovative batik with an unmistakably Malaysian identity. Like *songket*, batik has successfully made the transition to haute couture textiles (high fashion textiles). For me, today's artisans need therefore to explore new dimensions in existing traditions and develop an acute sense of perception to the needs of the consumers in the overseas market.

Inspiration from Nature

Malaysian designs are derived from man's inspiration and nature's wonders. Batik expresses cultural, aesthetic, intellectual sensitivity and the way of life. Most of the design and motifs in Malay traditional arts and crafts are predominantly from nature, taken from exotic tropical rainforests of floral and vegetal and also inspired by the assimilation of many cultures. In the old days, people believed that man and nature had a strong relationship. For example, among the Iban of Sarawak, most blanket weaving, *pua kumbu*, makes use of natural sources: cotton yarns and natural dyestuff taken from trees and roots. Even the motifs on the *pua kumbu* reflect the beliefs and values of the Iban people. In *tekat* most of the designs comprise flowers and leaves. Most of the designs strongly relate to Islamic values; for example, the motif of *awan larat*, means 'protacted' or trailing clouds. It is designed in such a way that it is almost impossible to trace where the designs begin and end. It seems like the whole design is connected in peace and harmony. Floral motifs such as *bunga cempaka* (*Michelia champaca*), *bunga cengkih* (cloves), *bunga anjung* (*Mimusops elengi*, a tiny sweet-scented flower), *bunga manggis* (mangosteen fruit), *pucuk rebung* (bamboo shoot) are popularly used in textile design in Malaysia.

Geometric designs include stripes, diagonal lines, square patterns, chequerboard, rectilineal lines or Dongson-like S and double S shapes. Normally, animal motifs are rare as compared to plant, flower and geometric motifs. Most of the animal motifs have been stylized in abstract or in symbolic form.

Traditional Design in Contemporary Batik

Looking back at the batik industries here in Malaysia, makes me realize that batik is not just a piece of fabric that you can simply design and sell it for any price you like. Batik has its own unique approach, not only in designs but also in different techniques, either using the previous methods or using high technology of computer-aided design. Even though it has become mass produced in most of the country, the quality has to be there, so that you can present your own cultural identity in different way.

As a textile designer, I am aware that batik designs represent cultural identity. It is important to identify ourselves with our civilization and hoping that the art and crafts of Malaysia can be presented in different styles in textiles. For example, batik inspiration from the Malay Perak embroidery. What inspires me much in this craft is that the beauty of the designs in Malay *tekat* are always limited.

Malay Perak embroidery or *tekat* is well known among the upper class of Malay societies, especially royalty. Court costumes for men and women are richly encrusted with stitchery worked on silk, gold and silver threads. In *tekat* embroidery, gold thread is embroidered on a base of velvet fabric. The technique of embroidering gold thread and gold paper appliquéd in relief is still practised today using silk or georgette in Perak, Kelantan, Terengganu and Sarawak.

Though it is an integral part of royal pomp, splendour and wedding ceremonies, the art of embroidery has since spread beyond the preserve of royalty and is now enjoyed by the common people, especially among the textile and fashion designers. One of the local Malaysian designers, Eric Tho, says: 'batik has its own traditional aesthetic values and should expose more in the future. The richness of our arts and crafts should become our inspirations in designing textiles or fashions.' I believe that we should go and search our own history, original culture and traditions so that we would really appreciate it. I think we can achieve a tremendous success to create our culture identity with the combination of past and future in designing new dimensions of batik.

In Malaysia, the process of looking into the nation's cultural heritage for inspiration had already begun with an interesting results. For example, the logo of the national airline, MAS, was adopted from the Kelantan kite, *wau bulan*. It is a symbol of cultural tradition which would fly for all time.

Normally, embroidery is limited to decorating functional articles, like clothing and furnishings. Great care and attention is also required to maintain and preserve an embroidered item, and this makes *tekat* pieces impractical for everyday use. I try to bring out the *tekat* embroidery in batik as motifs, so that people can enjoy wearing *tekat* motifs casually and freely. I want to show that not only are *tekat* motifs unique, but the can also be expressed freely in contemporary batik to create new images with a distinct Malaysian identity. Ethnic identity runs deep and will continue to translate itself into what we can see and appreciate.

New Dimension In Batik Production

The twentieth century, especially in recent decades, has been an unprecedented period in science and technology. As Malaysia advances technologically, the need increases for labour that 'works smart', creatively and innovatively. In particular, the rapid advancement in textiles technology has allowed more visually-oriented approaches in education and industry. As a result, the process of designing and producing will be much more faster. For example, the Integrated Computer Aided *Tjanting* System (ICATS) that has been designed by Standards & Industrial Research of Malaysia (SIRIM). It is a unified batik *tjanting* (*canting*) process in which soft- and hardware are designed, developed and controlled by machine. The system coordinates the various parameters including wax/resin temperature. These machines are able to operate using graphic software and to apply the designs automatically onto the fabric. It is much easier to develop and change the designs in terms of size and repetition.

This process is constructed primarily to enhance the speed and quality of the *canting*. It has been demonstrated that the combination of CAD/CAM technology and batik designer's skill are capable of speeding up the *canting* process while retaining the artistic value of the pattern. It increases productivity by eliminating the need for skilled labour. Skilled batik artists are thus able to devote more time to the hand-painting process. The computer is just the tool by which we can improve our production without taking a long time. As a result, the designer is the one who is totally involved in making things happen. Improvements include greater flexibility, speed, robustness and more adaptability in meeting the changing requirements of customers.

Conclusion

In the twenty-first century, people are lilkely to become more global in terms of product consumption, stimulating the use of high technology in many aspects of designing and producing. Batik in Malaysia is entering a new era of rapid and unprecedented change in the development of vibrant, energetic styles and high technology.

The non-Western world with its many diverse cultures is a seed bank from which textile designers will continue to draw inspiration for the foreseeable future. Textiles for all practical purposes are readily available nowadays. It has become high fashion in the global market. There is a need to adapt contemporary designs which give high value added and market potential.

The increasing accessibility of developing countries, especially coming at time of growing ecological awareness, gives the idea for designers to provoke a new trends in designing. The combination of technology and historical values creates new forms of thought, as societies change and aspire to a better life.

References

Arney, S. (1987), *Malaysian Batik: Creating New Traditions*, Kuala Lumpur, Malaysian Handicraft Development Cooperation.

Chin, L. (1980), *Cultural Heritage of Sarawak*, Kuching, Sarawak Museum.

Fraser-Lu, S. (1988), *Hand-woven Textiles of South-East Asia*, Singapore, Oxford University Press.

Norwani Mohd, N. (1989), *Malaysian Songket*, Kuala Lumpur, Dewan Bahasa dan Pustaka.

Ong, E. (1989), *Pua:Iban Weavings of Sarawak*, Kuching, Society Atelier Sarawak.

Peacock, B.A.V. (1981), *Malaysian Traditional Crafts*, Urban Council.

Selvanayagam, G. I. (1990), *Songket: Malaysia's Woven Treasure*, Singapore, Oxford University Press.

Siti Zainon, I. (1986), *Rekabentuk Kraftangan Melayu Tradisi*, Kuala Lumpur, Dewan Bahasa dan Pustaka.

Siti Zainon, I. and Kedit, V.A. (1994),*The Craft Of Malaysia*, Singapore, Tien Wah Press.

16 The Art of Batik in Sri Lanka

SIVA OBEYESEKERE

There is a long history attached to the art of batik in Sri Lanka. Although the ancient origins of this craft cannot be precisely established, it is possible that certain crafts were introduced to Sri Lanka from Java by the Dutch when they occupied the Maritime Provinces. The Dutch occupied Sri Lanka – then known as Ceylon – from 1657 to 1815. The Dutch East India Company carried this cloth, together with spices and fruits, to other colonies. Fruits like mangosteen, rambuttan and durian too were introduced to Ceylon through these traders.

Its name, derived from 'tik', signifies marking with spots and dots, and depicting cracks in the design on the cloth. The art of batik retains its original from and technique, except in materials and design.

In a country whose culture goes back 2,500 years it was an old custom to decorate places of worship with flags and banners made from batik cloth decorated with religious motifs. During the long period of colonial domination, batik art, like many other crafts, was dying out. Due to the efforts to revive and develop the traditional crafts of Sri Lanka after independence, batik's resurgence is one important craft that has brought forth the many artistic talents of Sri Lankans in our beautiful country.

Batik clothes are used by men and women as a form of dress – an old fashion that continues to the present. In early times government officials wore a batik cloth, known as a *Somana*, from waist to heels, while women wore batik saris richly decorated with colourful designs. Wall paintings in ancient Buddhist temples and some exhibits at the Colombo National Museum portray some of these *Somana* clothes. The material is very beautiful, rich in colour and intricate in design.

In ancient times, the natural dyes for batik printing were prepared from local materials. Beeswax was used for the resist material. These traditional colours were fast, and even today they have not faded and retain their original colour and texture. However, today the availability of a wide range of modern chemical dyes with more choice of colour has provided the Sri Lankan batik artist scope to add this exciting dimension to their technique.

Sri Lankan batik art expression ranges from its roots in Sinhala and Tamil legend, traditional motifs and themes through to scenes of everyday village life, into the flora and fauna, the environment, the sunrise and sunsets, the beaches and seas, which adorn and surround this island. The increasing use of modernistic interpretations of these many themes, the flair for bold use of colour threads and variety of stitches to highlight designs, and the breath of freshness is shown in contemporary products. Batik craftspeople have now developed their art to reflect their own indigenous designs and to distinguish their work from batiks of other countries.

Today batik printing is a flourishing cottage industry, mainly due to the increase in the tourist industry. Batik cloth is used for producing shirts, sarongs, saris, dresses, and a vast variety of ready-made clothes. Household linen in the form of bedspreads, tablecloths, cushion covers, place mats and numerous other items in this range are produced and used by foreigners and islanders alike.

The main feature of Sri Lankan batik is that it is handmade at every step of production. Batik articles are produced in large quantities due to the increasing demand, but the art of the crafts has never lost their quality and beauty. Batik has continued to retain the individuality of the artist and craftsman.

The Batik Art Method

Batik, in its genuine hand-worked form as practised in Sri Lanka, is the art of painting through the technique of wax resist dyeing. In the first stages the artist sketches the design on cotton or occasionally silk cloth (normally white) and the area that is to retain the original colour of the material is covered with wax to resist the seeping of the dyes. The wax is applied in a molten state using an instrument (the *canting*) which is a copper crucible with a thin curved tubular spout. The tool controls the flow of wax for precision of application. A combination of paraffin and beeswax is used for waxing. The cloth is then dyed in the lightest of the colours to be used in the design and subsequently given a further waxing to retain this colour in the final design. This process is repeated as many times as there are different colours in the design, with colours following each other in ascending order of darkness. This process of placing the material in dye baths is called the *cold method*. After this wax is removed by repeated immersion of the cloth in boiling water, which also ensures colourfastness in the final product. The more complex designs may require 50 to 100 hours of painstaking effort to complete the process.

The final batik is distinguished from other forms of painting on cloth by virtue of the fact that both sides of the cloth should carry an equally good image. However, the option of reversing the design , which may be preferable in the eyes of some beholders, is always available. The batiks produced demonstrate many options. Some batik artists sign their work, especially wallhangings.

For the Future

For the past 30 years batik craft has staged a great revival and is a very large cottage industry, providing employment to many rural people. To encourage new designs and quality products, batik craftsmen have to be competent artists and have a good sense of colour. Batik makers have to improve their knowledge to produce and present their work for contemporary society, while retaining, when necessary, some of its traditional values. Designs have to meet the fashions of the international market, and the presentation of the finished product needs to be 'eye-catching'. Batik has to be adapted to global markets, while retaining its unique Sri Lankan characteristics.

17 Persistence of Traditional Dress in Mesoamerica

CAROL HAYMAN

Traditional dress is embedded in the political, religious, social and ideological matrix of a society. Costume has three key functions: a) it is a marker of community affiliation and social identity; b) it asserts cultural and individual values; and c) it maintains a distance from the dominant culture. The wearer of traditional costume rejects the dominant cultural values.

As a marker of community affiliation costume provides an outward manifestation of group identity. The clothing expresses a particular ethnicity and a specific community locality. The style of the costume is prescribed by the community in forms that are determined by the traditions of the community. It shifts the focus from the individual to the community.

An example of how this is evident can be seen in the Zapatista communities of Chiapas in southern Mexico. Their desire for autonomy and self-determination is reflected in their social traditions, including the traditional clothing of women. Their traditional dress communicates that the wearers hold a certain world view that permeates their way of life, a world view common to other wearers of traditional dress though belonging to different ethnic groups.

When a community with a different world view from that of the national culture feels threatened, it employs strategies for preserving its identity. Costume is one mechanism for maintaining social distance from the dominant culture. Clothing symbolically designates the community's separateness from the outside world. The wearer maintains his or her own cultural values and rejects those of the dominant culture. Costume also emphasizes the separateness and uniqueness of each community, in contrast to neighbouring communities.

In a community with inner integrity, one that is in relative equilibrium, one can expect to see a limited range of symbolism and styles acceptable in that community. Colour choice and arrangement, shape and structure, are all dictated by community recognized tastes. The techniques and decorations of traditional clothing form a continuity with the past and with surrounding communities with similar, yet different, costumes. The continuity serves to reinforce a feeling of belonging to a community. The costume is a reflection of an individual's affiliation indicating conforming participation in a unified community. Its purpose is to 'deindividualize the individual' (Yoder, 1950, pp. 308, 318). The individual is submerged in the group personality of the village.

Traditional dress functions in a similar way to language. The community language or dialect, different from that spoken by the outside world, isolates that community, yet connects it to other communities that speak the same language. Costume isolates a community while connecting the members wearing it. People can identify the home community of an individual in the market by his or her garments. Clothing reinforces commitment to common values through the expectations of certain behaviour associated with style of dress.

Traditional clothing, called *traje* in Mesoamerica, carries an ethnic code woven into the fabric. The shape, the length, the colours, the decorative designs and the way it is worn all transmit information. Other members of the wearer's community instinctively understand the information. The costume announces that the wearer subscribes to and supports the values of her community. People outside the group boundary defined by the costume also understand the ethnic code: they may be wearers of a different traditional costume themselves.

Conformity in clothing helps maintain adherence to the accepted pattern of behaviour. If a community boundary is strongly defined through ethnic markers such as clothing and language, the existing social order is likely to be supported by strong rationalizations and emotional commitments. The members of a community will have come to the conclusion that their way of life is the best. Such a system is likely to be self-consciously resistant to alteration. It is organized defensively as a result of challenges in the past (Social Science Research Council, 1954, p. 977).

Commitment to traditional forms is a defensive response to outside threats. Ethnic traits are often intensified through retrenchment as a result of intimidation from the dominant culture. Change is often viewed as a potential threat to life and property. Interest in protecting a community from change and outside threats, encourages participation in political and religious affairs and commitment to the preservation of the existing social structure. Conformity in clothing is one mechanism for maintaining the stability of a culture through its solidarity function, the cohesion of in-group consciousness and a positive affirmation of heritage. People wearing the community costume tend to support the community values and share the same world view.

The modes of thought and behaviour of the subordinate class are visibly different from those of the dominant class. Basic values inform group self-images and influence their mutual relationships. The attitudes towards class are substantially different from each other. The dominant culture imposes its prejudices on the subordinate culture; the prejudices are implicit, reflecting the world view of the Ladino culture. Ladino culture is vertically structured and status conscious: it stresses competition and obedience at work, school and in the home (Colby, 1961, p. 774). Indian culture is conceived in more horizontal terms. Social hierarchy is de-emphasized. There is an attempt to maintain the attitude 'Here we are all equal' (Foster, 1967, p. 303; Redfield, 1950, p. 43). The emphasis is on age and etiquette. Authority is maintained through persuasion and influence (Colby and van den Berghe, 1961, p. 774). Continuity with the past is also an important aspect of wearing *traje*. Wearing it also shows respect, not only for one's parents and the traditional way of life, but also for one's ancestors. In the codices, we can see pre-Hispanic female figures weaving and spinning in the same positions and with the same tools that Indian women make traditional costumes today.

The pervasive dual divisions, between Ladino and Indian world views, extends into the realm of ideas. The unquestioned assumption of cultural superiority by Ladinos is because of their view of cultural differences in hierarchical terms. Ladinos learn in school of their relation to the state, the nation and to other nations, while Indians try to maintain a precarious harmony with the cosmological order (Collier, 1975, p. 12). The goal of Indian culture is to learn the actions which will prevent misfortune in a universe controlled by unseen, supernatural forces. The goal of Ladino culture is to manipulate the universe, to establish power over things, animals and other men (Gillin, 1968, p. 196).

Indian culture is undervalued by the dominant culture, romanticized on one hand and viewed as an impediment to progress on the other. Levi-Strauss (1974, p. 41) explains how dominant cultures view Indians in general:

> Primitive peoples are all in their different way, enemies of our society, which pretends to itself that it is investing them with nobility at the very time when it is completing their destruction, whereas it viewed them with terror and disgust when they were genuine adversaries.

The Mexican government supports archeology for tourist attractions and glorifies the Indian past while at the same time viewing the Indian of the present as a 'hopeless incubus on Mexico's economic and political progress'

(Colby, 1961, p. 465; Caso, 1958, p. 27). The Instituto Nacional Indigenista, argued that the nation's productivity will increase as the Indian's capability as a worker increases and the awareness of belonging to a vaster social organization is introduced.

The concept of the Indian as inherently inferior and incapable of advancement helps to rationalize the economic exploitation of the population for cheap labour and justifies the taking of indigenous land. An Indian community is often located on marginal land which in the past was of no interest to the outside world for exploitation. Population pressure and other factors have made Indian lands more attractive to outsiders. Where Indian lands are of better quality or contain valuable resources such as oil or timber, state policies often have the effect of reducing the amount of land traditionally available (Collier, 1975, p. 191).

In 1994 NAFTA, the North American Free Trade Agreement, was implemented between Canada, the USA and Mexico. This agreement is part of the neo-liberal economic transformation that opened Mexican and US borders to the world economy. It allows multinational corporations that have no loyalty to national identity or governments to move factories to low wage countries with lax environmental standards. For Mexico it means that cheap imports out-compete Mexican domestic products, ending jobs, suppressing wages and closing small businesses. For the former, it means that cheap American corn, produced with greater amounts of capital and better lands with higher yields, lowers the price that poor Mexican farmers can get for their crop. It was against this background that the Zapatistas rebelled against NAFTA and the Mexican government, motivated by the plea for indigenous autonomy. Protecting their traditions and communities is symbolized by their commitment to wearing *traje*.

The Indians lose in the competition for limited valuable resources. As a consequence of land problems and economic conflict, relationships with the outside world are often hostile, reinforcing their self-defined need to maintain a social distance. The Indians are politically impotent against the superior economic position of Ladinos (de la Fuente, 1975, p. 446). They are resentful and fearful because the authorities protect the exploiters (de la Fuente, 1968, p. 94). The police, the army and private militias allegedly assassinate indigenous leaders and local farmers with impunity.

Official policies are also designed to reduce cultural identity, ethnic pride and to increase assimilation and acculturation (Beals, 1968, p. 465). Politicians view the process of forcing Indians off traditional lands as aiding in the unification of an ethnically diverse nation. Because of the alienation of their lands, some Indian groups are forced out of their isolation from national life and into the lowest stratum of Ladino life. They give up their dress, language and rituals and once integrated they become invisible (Salomon Sitton, 1981, p. 5). Political submission is a result of intimidation through violence and murder.

The power structure prevents the weak from challenging the situation. Large landowners, timber companies and cattle barons have the resources to drive families off land to which they have claimed traditional rights. Corrupt officials in the Departamento Agrario are believed to dispossess Indians of their lands. State and federal machinery is mobilized against litigation by Indians trying to recover their traditional lands. The struggle to preserve Indian cultures and traditions presupposes a fight against nationalism (ibid., p. 8). National uniformity of culture is the official agenda (Beals, 1968, p. 465).

The attitude of subtle racism expresses admiration for the long dead 'noble savage' while offering a low opinion of and poor treatment of the 'corrupted' Indian of modern times. This is mirrored in the dilemma faced by the Indian whose traditional costume is of colonial origin. In the colonial period, the Indian clothing styles were sometimes prescribed by the Spanish colonial administration through formal orders. In some places the style has persisted for many generations and so is considered traditional and evidence of Indianness. A colonial survival can be viewed by Indians as a traditional cultural idem because they have made it their own. The idea of having a traditional costume, interpreted their way, and unique to their community, is more significant than the actual items of clothing or their origin. Its function is more important than its form.

The cultural trait of clothing cannot be isolated from other aspects of culture. Clothing as a symbol of identity is essential to the maintenance of the traditional Indian society, but is related to other factors. The secular and the sacred cannot be separated. Producing beautiful clothing, carrying layers of cultural meaning, is a religious act, a gift to the gods. The similarity of clothing in a community makes it a collective act. The act of dressing in *traje* is homage paid to the gods and the saints who protect the village.

In many areas, when asked why they wear a costume, the people will answer 'es costumbre'. It is the custom, because we always have, because it pleases our saint, or our parents told us we should (Cordry and Cordry, 1968, p. 11; Branstetter, 1974, p. 134; Sayer, 1985, p. 225). It is a source of social approval. These sentiments reveal an awareness of how costume is embedded in the matrix of their society. Part of the symbolic value of costume lies in the perception of it as being traditional and the appreciation of its role in the continuity of custom. 'It's always been this way,' they say, even though it has always been changing, possibly almost imperceptibly, over time.

Under the threat of external political and economic force, traditional societies often feel the need to retrench, to emphasize their own customs and values against those of the dominant national culture. The material symbol of clothing is one way of maintaining a public identity different from that of the dominant culture. Wearing traditional clothing is a socially sanctioned way of feeling superior. Maintaining traditional customs is a form of cultural resistance of the subordinate class to absorption by the hegemonic culture. Clothing as a cultural product places itself in opposition to the mainstream simply by its existence (Lombardi-Satriani, 1974, p. 104). Its contestative function is recognized by the dominant forces. It is implicit in the attitude that to be Indian is to be subversive. This attitude is exemplified by the actions of the Mexican military against the Zapatista Indian communities of Chiapas where small villages are surrounded by military encampments and subjected to daily military patrols.

The characteristics of the community act in concert to maintain the traditional way of life and social continuity. The world view and attitudes towards work of Indian society are completely different from the values of the national culture. The views are in conflict, both cultures morally certain that their way of life is correct and best. Members of each Indian community think their community is the best, but they are generally not in conflict with other communities, but rather with the dominant culture. This conflict causes the dominant class to discriminate against the subordinate class. Some individuals, or even entire communities, are unable to endure this discrimination and so change their ethnic identity. They become members of the dominant class, on the lowest rung of the social hierarchy.

In the national culture, Indian languages and culture are undervalued or even actively discouraged. Formal education is an effective means of national integration. Children are particularly sensitive to criticism and ridicule from authority figures such as teachers, and so gradually reject the ways of their parents. Some groups maintain their sense of value of their way of life. *Traje* helps sustain this sense of value by its symbolic meaning, but the traditional way of life cannot survive total disruption of a community through economic oppression, military repression or natural disasters.

The theory that the only way out of poverty for the Indians is through the abandonment of their few remaining cultural items in fallacious in two ways. First, it is based on the false premise that Indian culture contains few characteristics that are truly Indian. In fact, Indian culture is rich in features that have been symbolically reinterpreted to fit the needs of Indian society. The costume and religious ritual may have many elements of Spanish origin, but are now wholly integrated into the culture. Second, giving up cultural items will not make Indian life richer. The Indian will join Ladino society on its lowest, most poverty-stricken level. Many people do make this choice because they see it as a way of obtaining a better life for them and their children. Due to upheaval – social, political or natural – their traditional way of life can no longer sustain them.

That prosperity can only come to groups who abandon their traditional way of life is often contradicted by reality. Moderate prosperity can have the effect of allowing Indians to maintain their traditional way of life, including wearing their costume. When economic factors cause a drop in the standard of living to below the subsistence level, they have the effect of forcing Indians to become either Ladinoized *in situ* or to migrate to urban areas. Lack of ethnic pride is one result of a depressed economic condition. It is why some groups accept the ideology that Indians are inferior. Part of the population is not inferior, but excluded deliberately. They hover at the intermediate stop between Indian and Ladino, the second-rate Mexican who speaks Spanish poorly, the illiterate who has no culture. Loss of heritage causes self-depreciation in the spiral towards Ladinoization. On the one hand, Indians are encouraged to preserve their heritage, by collectors, for example, who urge them to reject acrylic yarn, while on the other, their experience suggests that to remain Indian is to accept inferior status.

Not all Indians see the transition to Ladino culture as an improvement in their situation. They recognize the value of retaining their traditional way of life, certain that it is the most appropriate for them and their children. This attitude helps them prevent cultural decay and individual demoralization in the face of cultural encroachment from the dominant society.

References

Beals, R. (1968), 'Notes on Acculturation' in Sol Tax (ed.), *Heritage of Conquest: The Ethnology of Middle America*, New York, Cooper Square Publishers Inc.

Branstetter, K.B. (1974), 'Tenejapans on Clothing and Vice Versa: The Social Significance of Clothing in a Mayan Community in Chiapas, Mexico', PhD dissertation.

Caso, Alfonso (1958), 'Ideals of an Action Program', *Human Organization*, 17, pp. 27–9.

Colby, B.N. and van den Berghe, P.L. (1961), *Ethnic Relations in Southeast Mexico.*

Collier, G. (1975), *Fields of the Tzotzil: The Ecological Bases of Tradition in Highland Chiapas*, Austin, University of Texas Press.

Cordry, D. and Cordry, D. (1968), *Mexican Indian Costumes*, Austin, University of Texas Press.

Foster, G.M. (1948), 'Empire's Children: The People of Tzintzuntzan' in Smithsonian Institute of Social Anthropology Publication No. 6 (1967), *Tzintzuntzan: Mexican Peasants in a Changing World*, Boston, Little, Brown & Co.

Friedlander, J. (1975), *Being Indian in Hueyapan: A Study of Forced Identity in Contemporary Mexico*, New York, St Martin's Press.

de la Fuente, J. (1968), 'Ethnic and Communal Relations' in Sol Tax (ed.), *Heritage of Conquest: The Ethnology of Middle America*, New York, Cooper Square Publishers Inc.

Gillin, J. (1968), 'Ethos and Cultural Aspects of Personality' in Sol Tax (ed.), *Heritage of Conquest: The Ethnology of Middle America*, New York, Cooper Square Publishers Inc.

Harris, M. (1964), *Patterns of Race in the Americas,* New York, Walker and Company.

Levi-Strauss, C. (1974), *Tristes Tropiques*, tr. John and Doreen Weightman, New York, Atheneum.

Lombardi-Satriani, L. (1974), 'Folklore as Culture of Contestation', *Journal of the Folklore Institute*, 11, pp. 99–121.

Redfield, R. (1950), *A Village that Chose Progress: Chan Kom Revisited*, Chicago, University of Chicago Press.

Salomon Sitton, N. (1981), 'Some Considerations of the Indirect and Controlled Acculturation in the Cora-Huichol Area', tr. Horacio Larrain, in Thomas B. Hinton and Phil C. Weigand (eds), *Themes of Indigenous Acculturation in Northwest Mexico*, Tucson, University of Arizona Press.

Sayer, C. (1985), *Costumes of Mexico*, Austin, University of Texas Press.

Social Science Research Council Summer Seminar on Acculturation (1954), 'Acculturation: An Exploratory Formulation', *American Anthropologist*, pp. 973–1002.

Yoder, D. (1972), 'Fold Costume' in Richard M. Dorson (ed.), *Folklore and Folklife: An Introduction*, Chicago, University of Chicago Press.

18 Batik Effect on Pure Woven Wool

DOUGAL PLEASANCE

Introduction and Background

Traditional hand-crafted batik effects are synonymous with Southeast Asia and, in particular, Indonesia. The World Batik Conference and Exhibition in Yoghakarta, the world capital of batik, highlighted the many aspects which combine to give it its appeal and social and economic importance.

Historically, batik craftsmen used naturally occurring colouring materials, such as indigo and *soga*. However, current practitioners use naphthol dyestuffs with certain highlight colours produced with solubilized vat dyes (indigosols).

The base fabric has traditionally been cotton, woven to a range of weights, fineness and price. Attempts to broaden the base for marketing batik have seen the introduction of fabrics constructed from such fibres as rayon, silk and nylon.

Fine wool is synonymous with Australia, and the International Wool Secretariat (IWS) has long pursued the goal of combining the novel effects and artistry of the batik industry with the international fashion appeal of wool, in order to secure a prestige section of the world fashion apparel market.

In 1994 Yayasan Batik Indonesia asked IWS to provide lightweight pure wool fabrics, suitable for making into batik textiles and garments by local designers. The fabrics were for display at the APEC economic leaders meeting, held in Bogor during November of that year.

In 1995 a cooperative approach to solving some technical problems was initiated by IWS. Aided by the GOI Institute for Research and Development of the Handicraft and Batik Industries (IRDHBI), ICI and their local chemical agent, PT Galic Bina Mada, an innovative new process was developed and successfully demonstrated at IRDHBI in Yogyakarta in February 1997.

This chapter represents the collected thoughts of the many initiatives in Indonesian and Australian research into developing appropriate base cloths for batik. These initiatives include devising modified batik techniques which

162

could be accepted by traditional batik practitioners and which offers batik manufacturers the opportunity to supply high value-added wool fabrics and garments to the international market.

Strategic Considerations

Planning the development of a wool batik process included a full and frank consideration of all the perceived benefits and problems associated with such a concept. These are listed as follows. Firstly, the benefits:

1 *International fashion acceptance.* Wool is a prestige fibre highly sought after by the international fashion industry. International fashion also demands continuous development of new methods of colour, design and texture applications. The combination of wool with the exotic image of batik is an innovative concept with international appeal.

2 *Drape, handle and texture.* Wool has unique properties, which combine to produce garments which drape well on the body. They retain shape and appearance, and they feel and look good to wear.

3 *Comfort.* Wool's water vapour absorbing properties contribute to its great wearer comfort factor.

The problems to consider are:

1 *Winter fabric perception.* Wool has been used in northern hemisphere winters for so long that it is perceived to be a winter-only fibre and, by deduction, something too hot to wear in a hot climate. The 'Cool Wool' programme has successfully positioned wool as a trans-seasonal fibre and provided the technical specifications for wool fabrics suited for wear in tropical regions.

2 *Skin discomfort.* Lined tailored garments are trouble-free. However, knitwear and dresses touch the skin directly and skin irritation can occur if the skin is particularly sensitive and if the wool fibre in the garment is coarse. Whilst we have no control over individual wearers' skin sensitivity, we can control the wool by specifying that its mean fibre diameter will be less than a prescribed value. It is in this area that superfine Australian

merino wool makes its greatest contribution.

3 *Felting shrinkage.* Harsh washing causes (untreated) wool garments to felt and reduce unacceptably in size. It was accepted that wool fabric to be used for batik would need a treatment to make it fully washable.

4 *Incompatibility with the traditional batik process.* Wool is normally dyed in boiling water. A cold dyeing method is essential when produced wax-resist effects. Wool is damaged by the caustic alkali used in traditional batik dyebaths. Untreated wool felts in the washing routines used to remove wax from traditional cotton batiks. It was accepted that the wool batik method would have to differ from the traditional cotton batik method. However, the modifications would be kept to a minimum.

'Engineering' a Wool Batik Base Fabric

Wool fabrics can be 'engineered' to meet a wide range of consumer requirements. When the task is to produce a trans-seasonal lightweight washable woven fabric suitable for batik application, the following parameters need to be considered:

1 *Wool micron.* The fibre diameter of wool is expressed in microns (one micron is one millionth of a meter) and the average, or mean fibre diameter of wool is measured during testing and subsequent processing by air flow or laser scanning techniques. Superfine is a term applied to the finest wools, arbitrarily to 19.5 microns and finer. Australia produced 80–90 per cent of the world's superfine wool.

When considering the choice of mean fibre diameter, a reduction in micron will result in: improved ability to produce a very fine count of yarn; improved comfort factor; increased softness of handle in the resultant fabric (increased luxury); and a small reduction in durability in the resultant fabric (decreased serviceability). In general, however, the finer the wool the higher its price.

2 *Yarn twist and ply.* Traditional worsted fabrics generally have a two ply warp and a choice of single or two ply weft. Count limitations associated with worsted spinning permit the production of plain weave fabrics normally with a weight range of 160–200 gm^2.

Further reductions in fabric weight require single ply warps, which in turn necessitates warp sizing prior to weaving. When considering the choice of yarn twist, an increase in twist factor will result in: increased crispness in the fabric handle (this is also associated with a reduced feeling of softness and a 'dryer' handle); cooler, denser, less hairy fabric (this is the basic principle employed in the production of 'Cool Wool'); improved resistance to pilling; and a small increase in yarn strength.

3 *Fabric sett ~ fabric weight.* Fabric weight is determined by the combined effects of yarn count and the fabric sett expressed as warp and weft yarns per cm. Having selected the fabric weight range required, a consideration should be that an increase in fabric sett will: increase the stiffness and durability of the resultant fabric, and reduce the drape and flow of the resultant fabric.

Fabric Preparation

Prior to the application of wax-resist batik effects, woven wool fabric must be chlorinated and set flat. Chlorination is an essential prerequisite to ensure adequate dye uptake. The flat set is a vital component of the fabric finishing routine to ensure easy care properties in the finished garment.

Chlorination

Chlorine is an extremely reactive chemical which attacks the surface of wool fibres in a manner which simultaneously assists dye penetration, and reduces wool's felting properties. Because of the high reactivity, great care is required in order to apply it at the correct level evenly throughout the batch.

Chlorine gas applied at the 1 per cent level is adequate to assist dye uptake and control felting shrinkage. This application is regularly applied as the first half of the chlorine hercosett process for wool top. The great advantage of processing wool in top form is that any irregularities which may have occurred during this process are blended out during subsequent gilling and drawing operations. (The term 'wool top' describes a length of twistless wool sliver wound into a ball weighing from 6–12 kg. It is an intermediate stage when processing wool through to yarn in the worsted system.)

Fabric (or *piece*) application of chlorine is an alternative which is used when the existing wool fabric source has not been pre-chlorinated. We have accumulated much technical information on how to produce even piece chlorination treatments in order to increase the versatility of fabric supply. However, we continually aim for the best results, which are obtained by chlorinating the wool in top form.

We have introduced this technically demanding process into a section of the industry which is skilled and practised in the art. By so doing, we can confidently expect that the fabrics offered to the batik industry will accept the dyestuffs applied by the following methods and resist felting shrinkage during the wax removal and subsequent domestic laundering phases.

The Batik Process

The chlorinated wool fabric is subjected to the application of molten wax in patterns using traditional waxes and application techniques. The waxed wool fabric is then coloured by the urea cold pad batch technique, which involves padding the fabric in a cold dye liquor and storing the fabric wet for 24 hours. After the dyestuff has reacted with the wool, a series of wet treatments are applied to remove the excess dyes, chemicals and the wax, and to soften the fabric.

Colour Paste Recipes

The dyestuffs used for this process are categorised as 'cold dyeing reactive'. Developed originally and marketed internationally by ICI as the Procion MX, an Indonesian company, PT Galic Bina Mada now manufactures (with technical back-up from the UK) and markets the same product under the name Chloranyl MX. The product range includes: Chloranyl Yellow MX-4G, Yellow MX-3R, Yellow MX-4R, Red MX-8B, Red MX-5B, Blue MX-2 and Black MX-GD.

The following table gives the general recipe for making concentrated colour solutions and the colourless reduction required for making pale shades.

Ingredient	Colour	Reduction
Dye powder	3.0	0.0
Urea	30.0	30.0
Manutex RS	0.5	0.5
Water	66.5	69.5
Total	100.0	100.0

Colour Mixing Method

1 Measure urea and Manutex RS into a mixing bowl.

2 Add hot (65°C) water and stir until solution is complete.

3 Add dye powder and stir until solution is complete.

(In the preparation of colourless reduction, step 3 is omitted).

The technique recommended for compound shades is to mix the 3 per cent concentrated solutions produced in this matter rather than by the dry dye powder method.

When producing pale or pastel shades, the reduction of colour is achieved by adding increasing proportions of the colourless reduction. For example, a recipe may read as '7 Red MX-8B, 3 Blue MX-2G 1–5' which implies that 7 kg of Red MX-8B paste will be added to 3 kg of Blue MX-2G paste to obtain the correct hue (in this case a dark purple). Then 50 kg of colourless reduction will be added to obtain a total of 60 kg of colour paste at the required depth (in this case, a pale purple or lilac). This procedure is modified in proportion to the required amount.

Cold dyeing reactive dyes are intermixable in any proportion. Dyeing should take place within 24 hours of mixing. If kept in airtight containers, this time may be exceeded. However, re-testing should precede bulk production.

Colour Application

The colour is applied by 'slop padding' in a simple immersion bath. Padded fabric lengths are layered flat on a sheet of plastic. Successive lengths are laid on top of each other either until the task is completed or a further stack is required. The completed stack is covered with a sheet of airproof plastic to prevent drying, and left for 24 hours.

Removal of Wax, Spent Dyes and Chemicals

The following calculations have been based on treatments where the volume of water used is 20 times that of the weight of the wool (clean and dry) being washed.

1 Prepare the bath with 5 cc/l ammonia and 2 g/l Valsol DW at 50°C. Scour for 10 minutes, and then drop the wool into a cold water bath. Remove as much brittle wax as possible. Return the wool to the ammonia/Valsol DW bath for a further five minutes, and then return to the cold water. Try to remove more brittle wax.

2 Prepare a fresh bath; mix 2 g/l Lanaryl RK with 8 g/l white spirit; emulsify this mixture into the scouring bath, and then add 1 g/l ammonia; add the wool and raise the temperature to 85°C; hold for 15 minutes.

3 Add cold water to this bath (cold overflow rinse) and continue stirring until the overflow water is clear and the temperature is below 50°C.

4 Prepare a fresh bath with 2 g/l acetic acid and 0.35 g/l Valsoft BMS at 40°C.

5 Remove the wool, squeeze or spin to remove the excess water, and then dry.

Post-batik Processes

Wool fabric will require little or no pressing when dry when it has been effectively 'crabbed' set during preparation and where cold overflow rinsing is employed immediately after the 'boiling' treatments.

Traditional worsted finishing generally employs a steam decatising process to flatten the surface and create the sought-after smoothness of handle. Similar effects can be obtained with a steam press, as is used for knitwear. A steam iron may be used, or, as a last resort, a hot iron on a damp rag. The application of dry heat to wool is least effective.

Summary

The wool batik process differs from the traditional method used for cotton in the following ways: naphthol dyes (applied with caustic soda) have been replaced by chloranyl reactive dyes and a neutral fixative (urea); dye fixation involves storing the wet dye-impregnated fabric under plastic sheets for 24 hours prior to washing off; washing off is carried out at 85°C in the presence of emulsified white spirit.

The wool batik process is identical with the traditional method used for cotton in the following ways: the fabric is delivered to the batik industry in a 'prepared for processing' form; molten wax is applied in pattern form using *cantings* or copper stamps; colour solution is applied in simple immersion troughs; wax cracking techniques can be employed; wax is removed in hot water troughs and subsequently recovered; fabric can be rewaxed and recoloured; speciality highlight colours can be hand applied.

The wool batik process provides additional benefits in the following ways. First, the restricted colour range of naphthol dyestuffs is supplemented with solubilised vat dyes, which are very expensive. The cold dyeing reactive range covers the complete spectrum, is favourably priced, and the application method is highly suited to hand painted effects. Second, the wool batik process is also applicable to silk, which has a similar chemistry to wool.

After extensive trials at IRDHBI and several leading Indonesian batik houses, we believe that the new technology permits a successful combination of Indonesian batik artistry with fine Australian wool fabrics. Batik wool represents an important extension of this famous art form into new markets which could extend from upholstery to high fashion.

19 Batik on Paper Bi- and Tridimensional Structures

ALICIA FARKAS

Antecedent

Basically batik consists of drawing with wax on a white background, so a blockage is produced that prevents the tincture penetration when the piece is submerged in a colour bath. Additional blockages permit dyeing with new colours, which are superimposed, according a predetermined plan. This is characteristic of reserve techniques.

Cracks and splits that are produced in wax permit tincture penetration so that very thin coloured lines take shape. These fine lines are distinctive of batik and are known in French as *craquelé*.

In craft and artistic realizations, the textiles used as a base for batik are silk, wool, cotton and linen.

Since 1968, I have researched, applied and developed an artistic expression with this technique. By way of exhibitions, courses and conferences I have diffused and transfered batik techniques in Argentina and other countries in Latin America, and Spain.

In the 1980s an apogee of paper culture emerged that stimulated by ecological recycling trends and vegetable species conservation. This was combined with a notable increase of craft papers manufacturing in mills, specifically for artistic use. This occurrence produced a huge reevaluation of this antique material, not only as a support, but as an artistic medium itself.

At the same time in the last decades weavers led the development of a new kind of 3-D art form that incorporates the use of various weaving techniques and fibre materials.

Investigative Development

Paper Selection

The purpose of the research is to explore the application of batik techniques to paper, and develop an appropriate methodology to obtain a result in which paper constitutes an artistic element in batik. The work begins with the selection of craft and industrial papers by means of analysis and reports, beginning with those of daily use to the most sophisticated for artistic use. The purpose of the investigation may be to create in middle schools and colleges an awareness of batik's craft and artistic applications.

The papers' aptitudes were checked with regard to mechanical resistance, elasticity, fibre type, extension and orientation, acidity gradation (PH), saturation grade, liquid permeability, blockage, tincturation and wax removal.

The use of chemical products that could damage the paper or produce posterior degradation has to be eliminated because paper demands more precautions than textile fibres in processing.

The following industrial papers were experimented on: Craf, Canson, Obra, Witcel, Multiwitcel, Conqueror and banana. Handicraft papers included papyrus, garlic, palo borracho and cotton manufactured in Cuyo National University. Cotton came from the mills 'El Manzano', 'La Villa' and 'Palermo' from Buenos Aires. Linen and abaca was from 'Guarro' mill, Spain. Pescia cotton paper from 'Magnani' mill, Italy, was used. Japanese fibre papers included Tableau, abaca hemp fibre; Sekishu, kozo fibre; Kitakata, gampi fibre.

In San Jose de Costa Rica University an interesting interchange was held in a Craft Paper Workshop with the Director, Dr Grace Herrera, and Professor Ileana Moya. Papers manufactured with local fibres were tested and technical information and a bibliography of papers from the United States of America and Japan was obtained. This meeting facilitated dissemination of Japan's papers for specific artistic use, which were highly suitable for the applications.

Contacts with Professor Laura Rudin, a researcher of Spiritu Santo University from Brazil, allowed access to banana fibre paper that she made and to an exchange of technical knowledge of handmade paper.

In Granada University, Spain, joint investigations were made with Professor Teresa Espejo, in the Cathedral of Restoration Paper and Textiles. Reports were made on the works executed during the investigation with the specialized and modern technological workshop equipment. This equipment allowed verification and corroboration of the research, with respect to the characteristics and kinds of papers selected. These interchanges provided knowledge of the wide spectrum of industrial and craft papers that meet the requirements for the application of batik.

Process

Blockage The blockage or 'resist' for the zones in which tincture penetration is not desired is made with wax and paraffin combined in different proportions, according to the flexibility desired for cracks, *craquelé*.

90% virgin wax	10% paraffin	avoid *craquelé*
75% virgin wax	25% paraffin	light *craquelé*
50% virgin wax	50% paraffin	intense *craquelé*
30% virgin wax	70% paraffin	excessive *craquelé*, not recommended

Wax and paraffin are fused in a temperature regulating device, and for its application vegetable bristle brushes are used. The *canting* is the traditional tool used in Javanese batik for linear and continuous designs. Wooden, rubber or metal stamps, *caps*, are used as seals, reproducing the wax impressions.

The blockage may also use guttapercha latex in a linear way to produce an excellent resist. With chemical adhesives it is possible to obtain the same effects to those obtained with guttapercha.

Tincturation The procedure for tincturation follows the order of chromatic values, from the higher to the lower. Each colour requires an immersion or superficial contact dyeing. The immersion time is very short, one or two minutes for light colours and absorbent papers, and this time will vary according to the desired fastness of colour.

Tests were made to detect blockage effectively, impermeable capacity, immersion resistance, saturation, colour fidelity in drying process and permanence under direct sunlight. It was necessary to obtain a deep knowledge of the characteristics of each paper, verifying that each technique had different effects in each fibre type.

Craft anilines manufactured with natural pigments in the Cuyo National University were tested. Other craft tinctures researched at San Jose of Costa Rica University included the pigments of the Native Boruca Reserve:

justicia tintorea	light blue, blue
tectona grandis	pink, purple
cucuma longa	yellow

theobroma cacao	coffee
bixa orellana	orange, red
hibiscus rosa sicencis	red, purple.

A wide spectrum of industrial anilines was tested and highly satisfactory results were obtained from anilines from Holland, Germany and England. An adequate proportion of aniline is half a gram for each litre of distillate water, though this may have variations according to the colour used. Recipients' dimensions must be adequate for papers to be dyed, avoiding excessive folds and paper damage.

To increase fibres' tincturation intensity, fixatives are used, taking the precaution of using those that will not alter the fibres' constitution (i.e. neutral acidity, 7–9 PH). Natural and Glauber salts were used as fixatives. The adequate proportion is to use 10 grams of fixative for each litre of water.

Wax removal The resist is removed by placing the piece between absorbent papers under the action of heat and vapour. The blocking elements, wax and paraffin, are transferred to the absorbent papers, leaving the colours free. This operation must be repeated, changing the absorbent papers, until the elimination of the resist.

Transference During the blockage removal process, if special quality absorbent papers are used, excess tincture mixed with the wax is produced. The papers obtained with this method have got excellent texture and it may be used as base for mixed methods in new works.

Bi- and Tridimensional Structures

Craft papers of palo borracho, papyrus, kozo, gampi and mitsumata fibres of low weight permit identical tincturation on both obverse and reverse faces. This produces a translucent decorative effect.

Translucent surfaces are reversible and allow access to the hidden face. The pieces may be displayed with a gap between them and the wall. The interplay of bi- and tridimensional planes facilitates creativity and playfulness. These objects are essentially founded in the tension that exists between the rectangular forms and more simple curved lines. These forms do not belong to the abstract geometric field, but are derived from organic movements.

These tridimensional objects are the product of a plastic artist, albeit using

Illustration 19.1   'El pajaro': batik on paper

Illustration 19.2   'El vuelo': batik on paper

Illustration 19.3   'El nido': batik on paper

Illustration 19.4 'El nido' (detail): batik on paper

painting styles with references to Asian culture. These works are the result of the artist's studies, aesthetic vision and cultural identity, realized in her way with reference to the art conceptual premises of our time, but in complete isolation from other batik makers.

Light utilization, natural or artificial, translates into a plastic component multiplying presentation possibilities. The spatial strategy embraces the area around the object, involving the spectator in a dialogue. It is an intermediation between the artist and the surrounding society.

Conclusion

Special characteristic papers are available for the application of the techniques of wax resist, tincturation and removal of batik. Papers made of fibres provide suitable bases for dyeing. The techniques of batik on paper allowed the maximum utilization of its perceptives and sensorial characteristics.

Transference of images during resist removal over special papers obtains works of a high artistic value. In this case the papers selected were of the highest quality and high cost. Best works were obtained with Molino Magnani Pescia Paper, Italy. Industrial manufactured papers of low cost are also appropriate.

The ductility, flexibility and transparency qualities of fibre supports the following characteristics of batik treatment: high chromatism, texture, craquelado and image reversibility. The latter permits the realization of tridimensional creations.

References

III Internationale Biennale der Papierkunst 1990, Leopold Hoesch Museum, Düren.
6th biennale d'art textile 1990–1991, Köln, Galerie Smend.
A Catalog of Artist Materials 95–96, Seattle, Daniel Smith Inc.
Belfer, N. (1992), *Batik and Tie Dye Techniques*, New York, Dover Publications Inc.
Eimert, D. (1994), *Paper Art, Geschichte der Papierkunst, History of Paper Art*, Köln, Wienand.
Malerie auf Papier, Paintings on Paper (1989), Stuttgart, Institut fur Auslands-beziehungen.
Orban, N. (ed.) (1991), *Fiberarts Design Book Four*, Asheville, NC, Lark.
Pirson, J-F. (1984), *La Structure et l'Objet (Essais, Expériences et Rapprochements)*, Liège, Pierre Mardaga.

20 Independent European Batik Technique on Easter Eggs

ANNEGRET HAAKE AND HANI WINOTOSASTRO

Introduction

During the last decades, the fashion of 'Easter markets' was developed in Central Europe. Based on several traditions of decorating Easter eggs, the old customs changed into artistic work.

Mostly, those customs originated from Eastern Europe and were of pre-Christian origin. The egg is known worldwide as a symbol of fertility and revival of nature. There were findings in ancient graves of egg-shaped sculptures as well as remainders of decorated natural eggs (Polak, 1980). In many cultures eggs are integrated into buildings to keep away evil spirits and call upon the blessings of the appropriate gods. The name 'Easter' or *Ostern*, in German, is derived from *Ostara*, the goddess of the revival of nature, light and springtime. Her emblem was a rabbit, a sign of fertility.

The egg as a symbol of revival fitted also into the spiritual content of Christian resurrection. Eggs were part of the rent which farmers had to pay to their landlord or church at Easter. It was handed down from the Middle Ages that godparents and godchildren exchanged decorated eggs during Easter, but nothing was said about the method of decoration. Dyeing was one of the techniques which are mentioned in old texts. The dyestuff consisted of extractions of plants.

In European folk culture, several batik reservation techniques on eggs are used. According to tradition, they differ in the method of wax application as well as in decorative motifs. Often, the motifs were adapted from the traditional ornamentation of national costumes or rural buildings which contain Christian as well as pre-Christian symbols. Together with textiles for personal use, batik eggs were meaningful Easter presents given by young girls to their suitors. All Easter eggs had to be complete, which means they were not emptied. Empty eggshells symbolized death, sterility and uselessness. It is significant that rural Easter traditions show great similarity, though it is hard to believe that there existed direct contacts between Eastern European and Hessian villages centuries ago (Bott, 1979; Grein, 1976; Skopová, 1995). The decoration of eggs is part of the European cultural continuum

Nowadays, batik eggs and eggshells with traditional patterns are still made for a growing number of collectors. Beside that, there are many new creations based on the traditional techniques. I will introduce three main areas for traditional batik eggs according to their characteristic tools.

Main Areas of Easter Egg Traditions

European people have many pre-Christian traditions which still influence daily life. Rites of fertility and the casting out of winter during springtime are associated with Easter and show similarity from Ukraine to Hungary, Poland, Czechia, Slovakia and Germany. There are countless varieties of egg decoration from these areas but only three techniques of batik according to the tools for the application of wax.

Hessian 'Geschriebene Eier' (Written Eggs of Hessen)

Hessen is a German province around Frankfurt am Main. In the northern part there are a few villages which still keep old customs. Sometimes, elderly women wear traditional costumes on an everyday occasion. On special occasions such as folklore festivals young villagers also take out their heirloom costumes. One occasion is an Easter market, where up to 100 people show and sell Easter eggs which they have produced themselves. Beginning with Lent (approximately six weeks before Easter), there are Easter markets from Hamburg to Bern, even in small cities, and thousands of visitors travel to those places. In Hessen these egg markets started in the 1970s. Some people had the idea to collect money for the restoration of historic buildings by selling traditional Easter eggs and decorations. The idea spread and now one has the choice of many venues every weekend. The traditional and contemporary egg artists work during a great part of the year to prepare for the demands of the visitors; they are usually booked up for the whole 'egg season'. The latest development is that American groups plan to tour the European Easter markets. A meeting is going on in New Jersey, USA.

Formerly, Easter eggs had been decorated as presents for parents, god-parents and other family members. In general, they had to be red and, of course, not empty. Young girls gave them to their boyfriends. Eggs for their fiancé were accompanied by a handkerchief with red embroidery showing in one corner a heart, the year and the initials of the receiver. This custom became important during the last century when many people of that area were forced to find jobs far from at home in the mining areas of Ruhr and Sieg. The gift of decorated eggs in a handkerchief would always remind them of the girl waiting at home.

The typical batik decoration of Hessian eggs was an epigram, written in *Sütterlin* letters, surrounded by borders of bread and wine, hearts or other meaningful symbols. The wax was applied with the help of an old-fashioned steel pen which was dipped into a small amount of wax and heated over charcoal. Sometimes, a spoon was used as a pan. Besides the writing pen, the glass head of a pin (*Spennel*) was used as a drawing instrument. Beadlike motifs were grouped in circles or half-circles and mixed with pen-drawn motifs and texts. These so-called 'sun-wheels' are also common in most other 'egg-areas'.

The dyeing was very delicate because the dye solution was hot. As soon as the wax started running off the egg had to be removed from the dye bath. Usually, they were dipped into the dye with a spoon. The short contact with the hot dye bath is not enough to let the egg white coagulate; it will still remain raw. Another method is to slowly heat the dye bath together with the egg. The melting process is controlled by a piece of wax in the solution. Due to the dyeing method which dissolves the wax while dyeing, only a single colour is possible. With the exception of the method of wax heating, all other tools and techniques are still in use.

For the receivers of traditional eggs it was a matter of honour to keep the gift as long as possible. There was a saying: 'If the egg breaks, the love will end.' An intact egg will dry slowly until the yolk has become a stony ball which will rattle inside and easily destroy a well-kept old egg. Storing in grain regulates the drying process and avoids the stony yolk. Thus, full eggs can be kept for many years.

For the dyeing process, the empty eggs are filled with cold dye solution so that it sinks under the surface. Later it has to be dried completely before the wax is melted off, otherwise, the dye will spoil the pattern. A slightly better method is to use water for the filling and temporarily close the holes with wax. Some people make batik on the full egg and empty it after the dyeing. The decorating of Easter eggs was also a social event. The girls did the writing and dyeing, while the boys had to take care of the glowing charcoal. Old customs are still alive and collectors from all over the world have stimulated the production of eggshells carrying traditional designs and epigrams.

The documentation of old customs, epigrams and egg designs, as well as records of well-known 'egg writers', helps to maintain the tradition (Bott, 1979; Grein, 1976; Seim and Velte,1997).

Sorbian[1] *'Gequälte Eier' (Tortured Eggs of the Sorbic People Near Dresden)*

From this area two methods of egg decorating are known. In addition to scratching designs into the shells of pre-dyed eggs, multicoloured batik patterns are typical for the area around Bautzen/Lausitz. While the motifs of the incised designs are mostly floral, the batik eggs bear graphic patterns of triangles, diamonds, and drops which are sometimes combined with pinhead drawings like those of Hesse. Rows of triangles which are called 'wolf-teeth' surround the so-called 'sun wheels' and protect them against evil spirits.

Up to six or even more dyeings can be observed on Sorbian batik eggs. The application of wax can be compared to batik *cap*. Birds' feathers are cut into shapes and used as stamps. Due to numerous dyeings, the temperature of the dye bath has to be lukewarm. It may not reach the melting point of the wax mixture and is thus kept at around 40°C.

There exists a variation of batik eggs, namely the application of multi-coloured wax, *Bossiertechnik* (modelling technique). The wax, which is also applied with cut feathers and pin heads, remains on the undyed eggs.

The 'chemical variation' of scraped eggs, mentioned above, removes the dye partly by drawing with acid on the pre-dyed egg. Both types show floral designs which are transferred from 'blue print' patterns (Zaroba, 1996).

Blue prints are made by a batik *cap*-like method of textile patterning, in which the reservation paste, *Papp*, contains a mixture of chalk, copper sulphate, several other mineralic oxides and vegetable oils. It is applied on one side of the cloth only by wooden blocks. Needles are used to complete the design. The only dye used after the reservation print is indigo.

Until a few years ago, the Sorbic people of Eastern Germany had few opportunities to obtain chemical dyestuff and were forced to find substitutes. Copying ink produced a brilliant violet. From vegetable dyes, which were much more difficult to handle, the extract of boiled onion peel gave the best results. Dyes of rye, red beets or other vegetables gave rise to pale shades, only. In the 1990s the Sorbs use tempered solutions of direct dye in addition to vinegar and obtain deep shades of red, blue, green or black.

Ukrainian *'Pysanky' (Written Eggs from Ukraine and Moldavia)*

The third area with Easter eggs of outstanding beauty comprises the Ukraine and the neighbouring states. Here, one can find fine ornaments on eggs which may be traced back to pre-Christian times (before 900 AD). The designs contain ornaments of plants and animals, crossing and different wave lines, triangles, squares, diamonds and stars which are arranged in bands or plane patterns. All motifs have symbolic meanings which were adopted by the Christian faith. Even the colours have symbolic content. For example, *red* is a favourite colour on eggs and indicates love and happiness; *white* means purity and innocence; *black* is the colour of death and remembrance of the dead; *green* symbolizes hope and innocence (Perchyshyn, 1995; Woloch Vaugn, 1982).

Decorated eggs, *pysanky*, were exchanged between family members and friends at Easter. It is said that people were keen to present the most beautiful eggs in their communities. The waxing is executed with a special instrument, developed in the Ukraine. It is called *kistka* and consists of a copper tube attached to a stick. The designs could also be found as embroidery patterns on traditional dress and textiles, and for house and church decoration. The neatly decorated eggs symbolically conveyed good wishes and were not made for consumption. For the latter purpose, hard-boiled unicoloured eggs were prepared. The name of this type, *krashanka* (plural: *krashanky*), is derived from the Ukrainian/Russian words for *colour, red* and *beautiful*. The *krashanky* were part of the food which was taken to the Easter service at church to be blessed. After strict fasting during Lent, blessed meat, eggs and dairy products were consumed for breakfast on Easter Sunday. The blessed food was distributed by the head of the family so that each member got his or her share. After breakfast, many games using *krashanky* have been recorded, which are also known in other 'egg-areas'. All the games are based on defeating others by destroying their *krashanka* while keeping one's own. The damaged egg will be given to the victor, who has to consume all of it.

Due to the political upheavals following World War I, many Easter customs appeared to have been forgotten in Ukraine, though Easter eggs continued to be made by several artists. One of them, Olga Bujuleva, became well-known in the German Easter markets. She allegedly makes the finest batik on eggs. The art of *pysanky* and the accompanying old customs have, however, survived in the USA and Canada, and were passed down from emigrants at the end of the nineteenth century. In 1974, the 'Largest Easter Egg of the World' was erected in Vegreville near Edmonton in honour of centennial celebrations of the Royal Canadian Mounted Police, the guardians of harmony and peace.

The aluminium construction is more than 10 metres high. Its design carries the symbols of life, good fortune, faith in the ancestors, eternity, rich harvests, and defence. The colours bronze, silver, and gold indicate prosperity. Original eggs are sold as souvenirs.

Varieties of Batik Technique

There are several methods of decorating eggs with batik. One procedure involves wax drawing on pre-dyed eggs which are then put in acid to remove the dye from uncovered spots. Formerly, people used lactic acid from *sauerkraut*, which has now been replaced by hydrochloric acid. It is practised in Czechia and the neighbouring territories of Germany. Sometimes, the eggs are waxed and dyed twice before they are put into the acid. Another method is the etching of undyed batik eggs until the acid engraves or even perforates the eggshell.

The Author's Batik Methods

Selecting the Eggs

Due to the enormous pressure which is applied to the eggshell during the emptying and dyeing processes, it is necessary to eliminate porous specimens. In front of a lamp, pores and cracks can be detected.

Emptying and Cleaning the Egg Hygienically

Two small holes are drilled into the clean egg after marking the spot with a household egg-picker. They must have intact edges to avoid cracking during the dyeing procedure. A high-speed drilling machine is the best tool for obtaining good results. The content of the egg is stirred with a needle, shaken and then, carefully blown out with the help of a ball. The egg white should not touch the outer shell, because it will hinder dyeing. The surface is cleaned by brushing it with concentrated detergent water and/or vinegar. In order to wash the egg's inside, it is filled, several times by an injection tube; after shaking, the solution is removed again with the aid of the ball.

First Waxing

After drying, the first waxing can be applied. The wax mixture must not melt while touching the egg during work. It has to be chosen to suit the climate. At first, both holes are closed, and then a pattern, which will remain white, is applied.

First Dyeing

For dyeing, the solution of direct dyestuff is poured into a glass with a screwed lid, which is filled up to two to three centimetres from the brim. To avoid undyed spots, the eggshell is forced under the surface of the dye bath with the help of one or more strainers and the screw cap. It is useful to soak the egg with the dye before closing the lid.

Second Waxing and Dyeing

After drying, the spots which will retain the first colour are covered with wax before the second dyeing. This second dyeing will nearly replace the first colour so that green, for example, appears next to red.

Finishing

To remove the wax, the eggshells are blown with hot air from a hair dryer, while the wax is distributed over the egg with kitchen paper or cloth. Thus, the colour is resistant against damp and does not need a coat of lacquer.

Batik Artists Well-known on Egg Markets

Some names may stand out among the great number of producers of traditional eggs. The recently deceased Auguste Mann and her daughters Maria Becker and Rita Gockel, as well as her sister Wilhelmine Becker, represent what might be called an 'egg-dynasty'. Maria Guntrum, Hedwig Hof, Maria Schick, Monika Riehl, Maria and Gertrud Zimmer and Maria Kräling are somehow related to this egg-dynasty in Hessen. Sorbic tradition is represented by Dorothea Solcia, Rainer Grosa and Werner Zaroba; the Ukrainian way is best demonstrated by Olga Bujulec.

Fine examples of contemporary batik and related techniques are made by

Ute Pohl and Brigitte Raab (known for graphic patterns related to Sorbic style), Roswitha Tröster (birds and poems), Ulla Melcher (multicoloured landscapes), Elli Bonik (Ukrainian-style batik with remaining wax), and Wolfgang Velte (etching of uncoloured eggs) as well as the Hungarian couple Ildicó Bodor and Josef Maleczki (multicoloured batik and perforation through etching).

Note

1   Sorbian is a Slavonic language.

References

Bott, I. (1979), *Ostereier-Malerei aus Mardorf und Erfurtshausen*, Langewiesche Nachf, Hans Köster KG, Königstein/Ts.

Grein, G.J. (1976), *Osterei und Osterbrauch in Hessen*, Sammlung zur Volkskunde in Hessen, Nr. 1. Museum Otzberg-Lengfeld.

Perchyshyn, N. (1995), *Ukrainian Easter Egg*, Design Book 3, Minneapolis, Ukrainian Gift Shop, Inc.

Polak, E. (1980), *Bunte Eier aus aller Welt*, Harenberg.

Seim, A. and Velte, W. (1997), *'Ich mache Dir das Ei recht schön ...'* – Zur *Ostereiermalerei im Amöneburger Becken*, D-35001 Marburg.

Serbske jutrowne jejka – Bunte Ostergrüße, 30 postcards (1996), Domowina-Verlag, Bautzen.

Skopová, K. (1995), *'Lidová tvorba'* (= Folkarts), German translation of chapter *'Ostern bei uns zu Haus'* by P. Houdková (1997), KADENCE, Pardubice/C.

Woloch Vaugn, M.A. (1983), *Ukrainian Easter – Traditions, Folk Customs, and Recipes*, Ukrainian Heritage Company, USA.

Zaroba, W. (1996), *The Making of Traditional Sorbic Easter Eggs*, Hoyerswerda, personal communication.

PART 4
BATIK CONSERVATION

21 Textile Exhibition Issues in Indonesian Museums

PUSPITASARI WIBISONO

Introduction

Textile collections should ideally be exhibited in accordance with museum principles. This means that everything for the exhibition is relevant to the purpose, vision and mission of the museum. In fact museums in Indonesia do not give much attention to the complete role of museums, because they still focus on improving the museum management, experts, skill, systems, and funding. Currently museums in Indonesia are increasing their awareness through developing museum staff and establishing the museum's public image to the public. ICOM, the International Council Organizations of Museums, provides a lot of information worldwide about museum development. It has been very helpful to museums in developing countries, particularly in Asia.

Museums began as human society's equivalent of cultural memory banks. In the later part of the twentieth century, museums have become multifaceted, multi-purposed, and multidimensional organizations. In the past few decades, museums have seen significant improvements in collection care and use, and in the fields of exhibition presentation and public programming. The field of exhibition development and preparation is a complex and demanding one.

The Museum Exhibition Mission

Museum means a dwelling for the Muses – a place for study, reflection and learning. Therefore, museum exhibitions help to define the institution: museums have the desire to 'sell' the institution, change attitudes, modify behaviour, and increase awareness. Museum exhibitions also have several other goals. These include:

- promoting community interest in the museum by offering alternative

189

leisure activities where individuals or groups may find worthwhile experiences;

- supporting the institution financially: exhibitions help the museum as a whole to justify its existence and its expectation for continued support;
- providing proof of responsible handling of collections if a donor wishes to give objects. Properly presented exhibitions confirm public trust in the museum as a place for conservation and careful preservation.

Exhibition Administration

Exhibitions require a large degree of management and administrative effort, in addition to the collection and production activities. Museum administrators deal with many matters. The administrative tasks relate directly to exhibition planning and production. These are: scheduling and contracting for exhibitions; contracting for services; production and resources management; and publicity and marketing.

Coordination and communication are essential. Every person involved in planning, managing, producing, and maintaining exhibitions must be aware of the project's progress. There are many matters to review when preparing to schedule exhibitions, including: available personnel; available time; accessible financial resources; prior commitments; other museum activities and projects; and the size of the galleries relative to the exhibition requirements. Managers also need to take into account national, religious, and local holidays and community events such as special days, sporting events, commercial sales and market days.

Publicity and Marketing

Coordinating museum activities with community events and interests can gain a museum much free publicity from the local media. Methods used to communicate what the museum is doing include brochures, pamphlets, mailers or fliers, newsletters, posters, announcements of openings or acquisitions, and catalogues. In order to know what strategies to employ the audience must be identified. Such events have the effect of making the museum a place for families and friends to gather, thereby creating a sense of well-being. Art fairs, demonstrations, astronomy viewing nights, holiday celebrations, open-house events – all can promote a feeling of belonging between a community and its museum. Publicity and marketing plans should be as much part of the planning process for an exhibition as the gallery plan.

The best promotion is to give visitors a memorable experience and let them talk about it. Promotional leaflets are a good way of ensuring that the visitors have something to carry away and show to people. If these leaflets are well written and are in keeping with the new look of the museum, then visitors will take away a good image.

Making Things Pleasant for Visitors and Extending Their Stay

Although many tourists visiting Jakarta and other cities usually visit at least one museum, unclear exhibitions are tiring. Most people cannot endure more than an hour at a time of looking at museum displays. Museum cafés and restaurants allow visitors to take a rest and return refreshed. Coffee shops run by the museum itself, or more commonly outside contractors, provide a share of the profits, thereby increasing the museum's revenue.

Exhibition Design

Designing exhibitions is the art and science of arranging the visual, spatial, and material elements of the museum environment into a composition that visitors move through. Design decisions should be deliberate and calculated, and executed to achieve maximum effect. Certain elements of design are fundamental to all visual arts. There are six main elements in designing exhibitions: value, colour, texture, balance, line, and shape. Value is the quality of lightness or darkness, having no reference specifically to colour. Values are associated with visual weight characteristics. For design purpose, values are important for emphasis, orientation, and attraction/repulsion. Judicious combinations of value with the other design elements can dramatically affect the visual impact. Value is controlled by pigment, surface treatment, and lighting.

Colour is an extensive subject. To attempt to cover all aspects of colour would be inappropriate in this context. Colour requires both the physical characteristics of light energy and the action of the human brain. Colours are perceived through the filter of perception and are ascribed meaning.

The human being is a central design consideration that influences and relates to all other composition-related factors. Human beings have only one archetype with minor variations in size, weight, features, and the like.

In design terms the lighting currently used could be adapted to avoid the expense of completely new display components. Preservation advice would, however, need to be given on whether the current showcases are of sufficient standard to display the collection without deterioration. Decisions on lighting should be made after consultation with conservators. The current use of natural light and the use of additional display lighting are all matters for conservation expertise.

Costume Exhibition Concepts

There has been a tremendous increase in the awareness of Indonesian textiles as a result of numerous exhibitions and publications in recent years. Indonesian textiles are visually exciting whether displayed flat or draped. There has also been an increase in exhibitions of gold jewellery and other art objects. However textiles were worn as a part of a costume with jewellery and other accessories or used for purposes such as decorations for ceremonies or markers. Often various textiles were combined in a costumes. Displaying the whole costume may be a more stimulating experience for the senses than just examining the textile component.

The purpose of the exhibition is to show the total costumes of various cultures in the archipelago. Special textiles and costumes are worn to denote status, rank and life cycle ceremonies such as weddings and community celebrations. Sumptuary laws often regulated dress codes within Javanese and other courts. Costumes and textiles conveyed the status, position, and gender of the wearer.

The Storyline

The storyline comprises a document that serves design and production by providing the framework for the educational content of the exhibition. The storyline involves several elements of: a narrative document, an outline of the exhibition, a list of titles, subtitles and text, and a list of collection objects. The process of storyline and text development begins at the point of origin for an exhibition idea. The conception of an idea carries with it an assumption that the conceiver has a notion, vaguely perhaps, of what the exhibition is to contain and what it is about. Before work begins it is necessary to determine how to communicate the exhibition's message, its interpretive strategy. This is the start of the storyline process.

Exhibition Development

The term 'exhibition' will be used to refer to a comprehensive grouping of all elements (including exhibits and displays) that form a complete public presentation of collections and information for public consumption. The type of display is first the object-oriented exhibition in which collections are central. Educational information is limited. The exhibition maker focuses on a direct aesthetic or a classification approach to presentation. Art is often presented in this way. The second kind of exhibition is concept-oriented, in which attention is focused on the message and the transfer of information rather than on the collections.

Exhibitions start as ideas that come from many sources: audience suggestions, board members or trustees, collections management personnel, community leaders, curators, current events, director, educators, staff and volunteers.

Exhibition development may divided as follows. First, there are product oriented activities which are centred on the collection objects and their interpretation. Second, there are management oriented tasks that focus on providing the resources and personnel necessary to completing the project. This leads to a schedule of exhibitions and the identification of potential or available resources.

According to the institution's mission, the museum should pay special attention to the following issues: What is the goal of an exhibition? Where is the exhibition to be held? Why are we holding the exhibition? Who is the exhibition for?

Education and the communicative aspects of both existing and new exhibitions should be given equal priority to the tasteful, secure and logical display of objects. Objects cannot speak for them selves. The beauty and technical interest can be admired but their significance in textile history and society cannot be understood without placing them in the appropriate historical and cultural context.

The exhibition of objects must be combined with high quality and well designed media, like graphics, models, audiovisual displays, and labelling. The support materials, such as sound effects, scents or odours, should be included where appropriate. When our senses of hearing and smell are stimulated the museum experience becomes more sensual and enjoyable.

The mission of exhibitions should emphasize the importance of developing publication programmes on each exhibition topic, and any other subject covered by the museum. The publication process may entail the production of postcards, leaflets, small samples, illustrations, pamphlets, scholarly books, and catalogues for specialists and non-specialists. Museum exhibitions must be easily understood, and therefore each object must be interpreted for all levels of visitors.

The purpose of an exhibition must be clarified, especially if displaying new findings. Museum activities are usually directed towards communication of displays for the public. If the displays are to acknowledge donors, then their participation needs to be clarified. There may be linked activities such as special collection exhibitions, book publishing, and performing arts.

The most important issue in organizing an exhibition is time. The organizers must choose a time for the exhibition that does not conflict with other important events such as holidays, and other national days.

The success for an exhibition would be commonly gauged from the quantity of visitors, especially at the opening. Sometimes, there are questions from the visitors concerning the next exhibition, especially with regard to opening dates.

Any type of exhibition must have a reason. There should be a clear link between organization and the purpose of the museum exhibition. Another consideration is the use of museum exhibitions to present new findings, often in cooperation with other institutions or professions.

Special exhibitions for children are prepared because the needs of children are usually part of the museum's mission. New interpretation methods take into account the educational backgrounds of visitors, especially with regard to children. Museums are not simply concerned with old objects, but may use contemporary artefacts where relevant.

Every museum needs friends, and increasingly museum management relies on the support of unpaid 'friends' groups who raise financial aid and support in kind. These groups are usually made up of people who have the time, interest and resources to invest in support of an institution and people. Jakarta's Textile Museum is a member of the Indonesian Textile Association, and the Wastaprema, an organization of textile enthusiasts. Both of them are very important for the existence of the museum. The Paramita Jaya group of more than 51 museums located in Jakarta acts as a liaison and communication body.

The Textile Museum of Jakarta collects textile objects ranging from traditional clothes up to modern costumes. It still has a limited number of objects. The collection is stored in a building which was not really constructed for a museum. The collection consist of weavings, embroidery, batik, *songket* looms, and costumes. It covers textiles from all Indonesian provinces. In 2000 the Jakarta Textile Museum plans to develop a neighbouring building. The collection will comprise costumes and accessories. Textiles from the ASEAN countries will also be housed in this building.

The Jakarta Textile Museum is frequently invited to hold an exhibition or textile display outside the museum. It could be either a temporary or portable exhibition. Some of the outside activities incude:

FKN (Festival Keraton (palace) in Cirebon) June 1997. A small display was placed in the front verandah of the Keraton Kasepuhan Cirebon. It shows 54 pieces of Cirebon batik. We displayed the objects on the *gawangan*, accompanied by a simple leaflet of the objects. It was a big risk that we presented the objects in the open space without security and air conditioning, but the public are very interested in such displays.

The Vatican Textile Display. In the residence of the Indonesian Republic in Vatican we showed 40 examples of textiles. All the objects were placed in an informal exhibition to accompany the programme women's organization intern of KBRI Vatikan. The display was followed by a lecture about the symbols of Garuda by a lecturer from the University of Indonesia, Jakarta.

The Aminef and HTI. A lecture on textile and costumes was held in the Textile Museum in cooperation with Aminef Usis Jakarta as part of the Fullbright Programme with HTI (Indonesian Textile Association). Mrs Pia Ali Syahbana, the leader of HTI, and Mrs Yudi Achyadi gave the lecture to the visitors.

Sarawak Textile Exhibitions and Seminar in 1987. The Textile Museum was invited to have a display of textiles. We displayed the textiles in a large space accompanied by leaflets for a short exhibition.

The Netherlands Textile Museum Tilburg. The exhibition entitled 'Batik Sukma Jawa', in cooperation with the Textile Museum Jakarta, displayed 86 pieces. It was accompanied by film, gamelan orchestra, and fashion show from the Iwan Tirta Collection.

The Jakarta Textile Museum also received invitations from other countries such as Singapore National Museum, and was involved in conservation training by Dr Mary Ballard from Smithsonian Institute Washington DC. Other international conferences were held in in Basel, Canberra, Kuala Lumpur, Jambi and Yogyakarta.

References

Achjadi, Judi (1997), *Textiles and Costumes, Lecturing Article for Indonesian Textile Society (HTI)*, Jakarta, Textile Museum.

Akram, Basrul (1997), *Usaha-usaha Inovasi Tata Pameran di Museum, Museografia*, Jakarta, Depdikbud.

Alt, M.B. (1982), 'A Cognitive Approach: Understanding the Behaviour of Museum Visitors', PhD dissertation, London, Institute of Education, University of London.

Berghuis, Paul (1995), *The Consequences of the Core, Concept and Guiding Principles*, The National Museum Project.

Brawne (1982), *The Museum Interior: Temporary and Permanent Display Techique*, London, Thames & Hudson.

Dawes (1960), *Fundamental of Exhibition Measurement*, New York, John Wiley and Sons.

Dean, David (1994), *Museum Exhibition: Theory and Practice*, New York, Routledge.

Gesche-Koning, Nichole (1996), *The Education Role of Museums toward the XXIst Century, Study Series,* France, ICOM-CECA (International Committee for Education and Cultural Action), pp. 4–5.

Grant, Alice (1996), *Museum, Information and Collaboration: Why a Single Standard Is Not Enough, Study Series*, France, ICOM-CIDOC (International Committee for Documentation), pp. 9–10.

Hein, Hilde (1990), *The Exploratorium: The Museums: As Laboratory*, Washington DC, Smithsonian Institution.

Loomis, Ross J. (1996), *Learning in Museums: Motivation, Control and Meaningfulness, Study Series*, France, ICOM-CECA (International Committee for Educational and Cultural Action) pp. 12–13.

Malaro, Marrie C. (1985), *A Legal Primer on Managing Museum Collections*, Washington DC, Smithsonian Institution.

Pearce, Susan (1989), *Museum Studies in Material Culture*, Washington DC, Smithsonian Institution.

Roberts, Andrew (1996), *The Museum Information Profession and CIDOC, Study Series*, France, ICOM-CIDOC (International Committee for Documentation), pp. 5–7.

Subagiyo, Puji Yosep (1995), *Konservasi Tekstil – Textile Conservation, Conservator's Manual*, Jakarta, Textile Museum.

Thompson, J.M.A. et al. (eds) (1984), *Manual of Curatorship, A Guide to Museum Practice*, London, Butterworths.

22 Batik Conservation in the Rotterdam Museum of Ethnology

LINDA HANSEN

Introduction

Collections form a museum's foundation, its core and primary reason for its existence. If there is no collection any more, a museum must close. Collections will always be a constant source of care. A lot of energy, money and work goes into delaying the natural process of decay as long as possible. But taking care also gives a lot of pleasure: in sensing special object, in doing research, in passing on knowledge about artefacts, and in the awareness that next generations will be able to enjoy these objects.

Within ethnographic collection textiles constitute a vulnerable group of objects. They require special treatment that can only be given by specialists. Conservation is the binding factor in preservation and management. This is also the case in the Rotterdam Museum of Ethnology (RME). The museum owns a fine textile collection, including examples from all of the world's civilizations and from almost every period in history. Among the 11,000 pieces are archeological fragments, carpets, silks, embroideries and costumes. Indonesian textiles predominate, and particularly well represented are batik cloths of Java, Madura and Jambi. The collection is stored in a general department that meets high conservation standards. It is looked after by a trained staff of a registrar, documentation specialist, a restorer, a curator and a conservator.

In this chapter I will focus on the batik heritage of the RME. First I will describe how the batik collection was built up and how dedicated directors were aware of the importance of conservation. Then I will go into the actual policy of conservation.

The Batik Collection of the RME

The RME was founded in 1884, more than 115 years ago. The present core collection of batik comprises dyed cloth from Indonesia, mainly from Java, but also from Madura and Jambi on East Sumatra. Some of the cloths are considerably old and date from the first half of the nineteenth century. The most recent acquisitions are brand new. Together, the approximately 800 batiks give a fairly good, though not comprehensive, view of what has been produced, especially between 1850–1990, at least in Java. As such they form a coherent whole.

The basis of the museum was the extensive and important Indonesian collection of the Dutch Missionary Society, which had been active in Indonesia since 1813. From missions throughout the archipelago (Java, Sulawesi and the Moluccas) the missionaries sent objects used by the indigenous peoples to Rotterdam, its headquarters. There the objects were put on show for the public, who were told the background stories by missionaries on leave in Holland. The original loan included 15 batik cloths. This was not much, but luckily a donation later on filled this lacuna very well.

A donation of about 100 batik cloth in 1884 formed the major cornerstone for the museum. It came from one Dr Elie van Rijckevorsel, a promotor of eminent importance for our museum. As a physicist he undertook a study journey to the Dutch Indies from 1873 until 1878. During his stay he collected ethnographic objects of which the Javanese batik cloths were highlights. They may be considered the core of the batik collection of the RME, and they distinguish themselves especially concerning the north coast (Pasisir) batik. De van Rijckevorsel's collection entered the museum's precise documentation on origins and the names of patterns. The cloths came from all the important batik places.

The next founding father was G.P. Rouffaer. In 1928 he donated to the museum a group of 11 batik cloths from Java. From 1885 (at the age of 25) to 1887 he stayed in Indonesia, where he became interested in the colonial situation. His standard work on batik art in the Dutch Indies and its history, which he wrote together with Dr J. Juynboll is well known among specialists. It was published in 1914. In 1898 he became assistant-secretary of the Royal Institute of Linguistics and Anthropology in The Hague. He wrote detailed studies on Javanese history, art and batik. The first volumes of his work on the art of batik appeared shortly afterwards, and was completed in 1914. This publication is still a precious work on account of Rouffaer's broad view. In 1928 he donated a large collection of textiles to the museum. This was to the benefit of the then director, Mr Nouhuys, who also showed great interest in Indonesian textiles. Of the 87 items 22 were batiks from Java, some of which are portrayed in his work *The Batik Art*.

In 1929 the museum obtained from Mr Mees, a well-respected businessman in the city of Rotterdam, 20 batik cloths of high quality from Java. Most of them had not been used. Probably he had bought them in the first place for commercial activities. From the 50 batiks he owned, the museum was able to make a choice of 20, linking up well with the existing collection.

In the 1930s the museum acquired from Mr Jacobson batik cloth from Java, dating from the last quarter of the nineteenth century and beginning of the twentieth century. These batiks had probably been worn by his family members.

In this period also the collection grew with 13 batiks from the Leiden Cotton Company, dating from the last quarter of the nineteenth century. These batiks were bought in order to deliver examples for the production of export-cloth for Java. This was also done by the Kralingen Cotton Printers from whom the museum possesses some sample books.

During and after World War II no interesting pieces found their way into the collection. Fifty-five new pieces are mentioned in the annual reports. And at the end of the 1960s a consignment of commercially made postwar batik from Java's north coast reached the museum. The reason why the museum collected these batiks, which were neither of good quality nor in good condition, probably explains something about the doubts within the museum; is there still a future for batik arts and crafts?

In 1961 the Rotterdam Museum Boijmans van Beuningen transferred ownership of about 30 textiles, with some 20 silk shoulder cloths from the north coast, which showed Chinese inspired *lok-can* patterns. Probably this transaction happened through mediation by the curator of Indonesian Collection, Mr Hurwitz, who was preparing an exhibition of Javanese batik. The exhibition was held in 1962 accompanied by a catalogue *Batik Art of Java*.

In the 1960s the batik collection was expanded through purchases of cloth produced after World War II and decorated by the use of *cap* and synthetic dyes in exuberant colours. The applied patterns and motifs reflect the postwar period in which there is room for innovation and modern creation. The museum also bought about 15 batiks of the Indo-European type dating from the second half of the nineteenth century. They were mostly of superb quality.

In 1965 Mrs Alit Djajasoebrata became curator of the Indonesian Department. Her dedication toward batik has resulted in bringing the collection to a homogeneous whole. In the beginning she focused her interest on the

Illustration 22.1   Labelled hipcloths (*kain panjang*) in the textile department, collected by E. van Rijckevorsel and G.P. Rouffaer, donated to the museum in 1884 and 1920

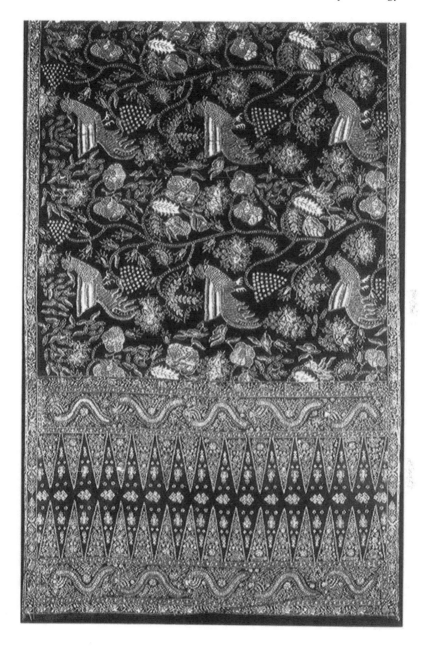

Illustration 22.2   **Tubular hipcloth (*sarung*), 112 x 206.5 cm, Semarang, North Coast (detail), collected by E. van Rijckevorsel, donated in 1884**

Illustration 22.3   Bedcover, 188 x 187 cm, Ungarang (near Semarang),
North Coast (detail), produced at the batik
workshop of Mrs von Franquemont before 1867,
collected by Rouffaer in 1900

Illustration 22.4   Silk hipcloths (*kain panjang*), 117 x 307 cm,
Surakarta (detail), produced at the batik workshop
of Mrs Meyer van den Berg c. 1900, signed H.
Meyer v.d. Berg

Illustration 22.5   Batik long sleeve men's shirts

recent developments in batik art. In 1968 she acquired seven contemporary batik cloths produced by Iwan Tirta. Together with Mr B. Sudibjo and Hardjono Gotikswan, Mr Tirta reintroduced old patterns, in order to put new life to batik. Especially noteworthy is a silk sarong with *lok-can* motifs, hand-made by Iwan Tirta, that helped to revive forgotten traditions of batik on silk. Through his initiative, the batiks of Trusmi in the neighbourhood of Cirebon were reborn.

Batik from Mrs Nora Yap and Mr Hardjonogoro were collected as well. Mr Harjon is known as a renewer in the Vorstenlandse batik and due to his merits was ennobled by the Susuhan of Surakarta Pakuwono XII. Mrs Djajasoebrata undertook several journeys to Java and Bali in order to fill in the gaps in the collection. Purchases were made at markets, in shops and batik workshops and sometimes from private persons. Batik cloths from Pekalongan, Banyumas, Garut, Tuban, Kedungwuni and Tasikmalaya found their way into the museum. Her trip in 1981 resulted in a book, *Flowers of the Universe*, and the batiks involved were on show in the museum in 1986 under the same title.

A remarkable asset to the collection was the purchase of a collection of silk *kain panjang*. This series was produced in the Indo-European batik workshop of Mrs Meyer van den Berg, and were all ordered by one person. The wearer was a lady of the Vorstenlandse aristocracy, married to a wealthy plantation owner of Indo-Chinese origin. These aspects are detectable in the textile through its patterns.

Mrs Djajasoebrata's latest acquisition for the museum was a collection of batik men's shirts which were made popular in the 1960s as formal clothing by the respected governor of Jakarta, Mr Ali Sadikin. His purpose was to emphasize Indonesian identity in modern men's costume and to support batik crafts. The idea soon became accepted in the whole archipelago. In the beginning the long-sleeved batik shirt was made from women's *kain panjang*; later on patterns were especially designed for men. This collection contains about 75 pieces, almost all of which were collected from charity societies. Together they give a good impression of what middle-aged men from high society used to wear in the 1980s and are still wearing in the 1990s in Java.

Conservation of the Batik Collection

Conservation of batik is a comprehensive task. Its organic basis (cotton, silk) makes batik cloth vulnerable: it faces different forms of damage, which can be divided into three categories.

Biological Damage

- Insect-infestations: moths attracted to dirt and food stains on the surface; cockroaches eating cloth and making large holes with their jaws, especially along the borders of folded batik; silverfish and ovenfish attracted by proteins.
- Mice: eat the fabric and use it as nest building material.
- Mould growth: white stains on the surface, due to too high RH. Mould spoors can develop easily at a humidity-level higher than 60 per cent and a temperature above 21° Celsius.
- Brown stains due to too high humidity level: these stains cannot be removed.

Chemical Damage

- Damage by light: discoloration caused by too intense and too long exposure to daylight and/or artificial light; it may also break down fibres, causing tears, rents and slits.
- Brittleness of fibres by dehydration: fibres may deteriorate, causing batik to tear or fall apart easily when handled. This is caused by too much heat and dryness in the fabric, which leads to cracks in the fabric.

Not all damage is visible right away, but fibres suffer a lot from fluctuations in climate; if this occurs too frequently batik eventually falls apart. When these fluctuations are avoided as much as possible the lifetime of batik cloth is prolonged.

Dye materials used in the batik may react chemically with the material and the environment. Black dye has an etching effect on the material, and the use of camphor to deter moths makes the fibre stiff and causes splits in the cloth. Acids released from cardboard and wooden cupboards also cause brittleness.

Damage Due to Using and Wearing

- Damage due to inaccurate handling in the museum, exhibition, storage on loan, etc.
- Tears, holes and rents caused by intensive wearing, but also dirt and stains.

The outcome is that a batik collection is constantly threatened by decay unless a museum has adequate conservation and restoration facilities. At present Dutch museums tend to give priority to passive conservation, which can be described as a set of measures and actions aimed at creating the best possible environment. The goal of p.c. is to guarantee the actual condition of an object and to prevent it from possible decay. Measurements and actions consist in the first place of climate control (temperature and humidity), lighting regulation and the prevention of air pollution, draughts, dirt, vermin, fire and burglary.

In view of its history of acquisition and its remarkable age, we may conclude that the batik collection is in a good state. One of the reasons is that the batik cloth itself was in good condition at the time it was acquired. However, the most important reason is the fact that museum's management throughout its history was always very dedicated to the well-being of its textile collection in general and to in Indonesia textiles specifically. This deep concern for the collection by directors – all male – is reflected in the annual reports of the museums. Actually a very concrete policy for 'passive conservation' was already present in 1885, and was thought of as a basic task of a museum. I will restrict myself to a few examples.

At the opening in 1885 the ballroom of the former yacht club on the River Maas was designated for the presentation of textiles. The public and the press were very enthusiastic. At first the batik cloths were probably either shown outside showcases, just like the displays at the Colonial Exhibition in Amsterdam, or in a diorama-like exhibition. The condition of the building was a constant concern of the Board. In 1900 necessary repairs were conducted and in 1902 the sun-blinds were replaced by cotton blinds, to enable enough light to enter but to stop harmful rays. Also a stove was installed to create a stable climate. It was considered essential to put the objects behind glass because of the harmful influence caused by moths, dust and smoke from steamboats.

New acquisitions were treated in a box filled with sulphuric-based fumigant. New showcases were purchased in 1903, which let in as little polluted air and light as possible.

In 1912 the museum was enlarged by a new floor on top of the building and the necessary precautions for the collection were taken. Paper covers were made to prevent dust from entering. The new textile gallery was opened to the public and the museum was able to display the collections in specially designed cupboards. Central heating and electric lighting were installed and even hydrometers. Because the collection was growing, especially the textiles of

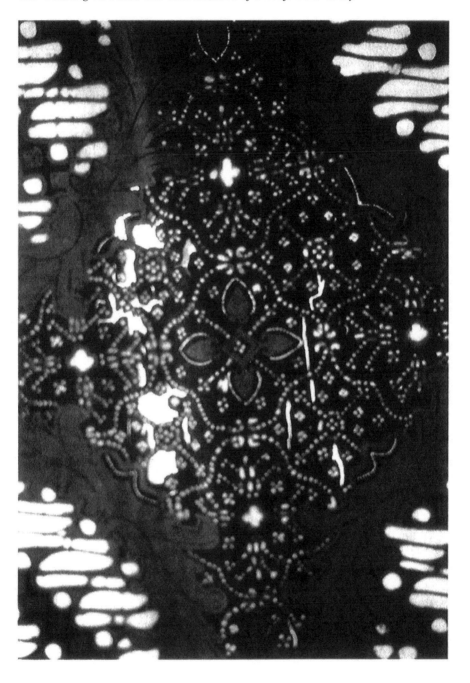

Illustration 22.6   Damage of black dye on batik cloth

Illustration 22.7   Damage caused by dirt

Illustration 22.8   **Showcases, Indonesian Department, Rotterdam Museum of Ethnology, 1903**

Illustration 22.9   Showcases protected from dust by paper covers during renovation of the museum in 1912

**Illustration 22.10a Wooden cupboards with rolled batik, left large
cloth and right small cloth, in the museum's storage
department**

Illustration 22.10b Wooden cupboards with rolled batik, left large
cloth and right small cloth, in the museum's storage
department

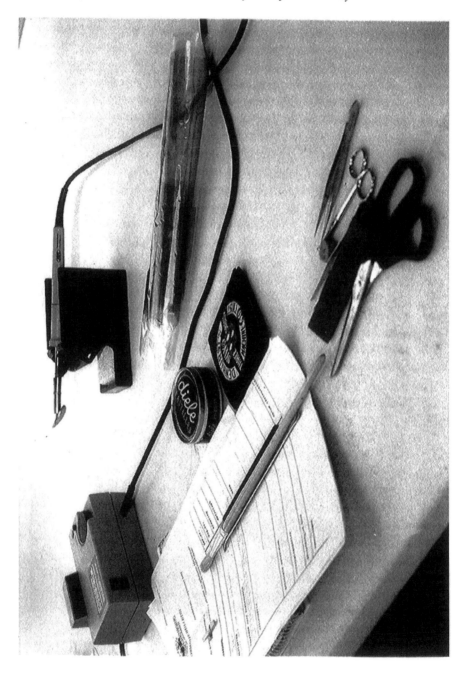

Illustration 22.11   Restoration equipment for treatment of batik cloth

Indonesia, the need for a storage department and study room was felt. At the outbreak of World War II, the museum secured a large part of the collection in safe places outside Rotterdam.

So this deep concern for the collections in the past enables us in the 1990s to enjoy the batik collection. Constant care is still needed. Today the collection is kept in a special department in the municipal storage building, situated 6 km away from the museum. Where the appropriate conservation conditions are maintained: a stable climate, with a constant temperature of 18°C and RH of 53 per cent. The batik *selendang, sarong kemben, kain panjang* and *dodot* are rolled over acid-free cardboard to prevent creases developing that will damage the cloth. The gold leaf on the *perada* cloth is protected by rolling in pieces of acid-free paper at the same time. The batik *kemeja* are hung in a cupboard, and are supported by coat hangers padded with cotton wool. When working with the textiles it is necessary to wear cotton gloves to prevent grease, dirt and acid from human skin from coming into contact with the batik. This storage area is cleaned regularly to prevent infestation. To restrict handling as much as possible the entire batik collection is recorded on colour-slides, which are scanned onto a computer system. These slides can be viewed on computer screens with the relevant documentation, so less handling is needed.

The collection is presented with great care. Batik cloth is preferably put in showcases, but this not always possible. When not in a case the batik should be on display for limited time, e.g. two months. Alternatively, one could change a piece during the duration of the exhibition. If batik is shown one should avoid attachment as much as possible. Hanging over a tube or laying flat may be a solution. Sewing will leave little holes in the material, which should be avoided as much as possible.

If the condition of the batik is poor, active conservation or even restoration is needed. The active conservation of batik consists of taking care of a cloth in such a way that it will not fall apart or allow tears and holes to widen. A sewing or gluing technique is used to treat problems. The treatment should not be applied to batik which was lined with a cloth in the 1970s. The function of batik has changed into a static two-dimensional piece instead of a three-dimensional piece of clothing. Ironing on silk or artificial gauze impregnated with a film of glue leaves the batik supple after treatment.

If old restorations exert too much tension on the cloth one may consider removing them; but normally these are not removed because they give information on its use or its wearer. Finally, the batiks are not washed. If stains cause problems the spot must be cleaned locally.

After 110 years the batik collection is still in a very fine condition thanks to the constant care and concern of museum staff. This involves an ongoing struggle against time and competition for funds which may be directed elsewhere in accordance with museum policy. Keeping the batiks in a good condition for future generations remains a central responsibility of the museum.

References

Annual Reports of the Museum of Ethnology for years 1885, 1887, 1890, 1899, 1901–06, 1908–10.

Djajasoebrata, A. (1992), unpublished catalogue of the batik collection, Rotterdam, Museum of Ethnology.

Hanssen, L. and van den Meiracker, K. (1997), *Theory and Practice of Museums' and Collections' Management*, Rotterdam, Museum of Ethnology.

23 The Batik Collection at the Tropenmuseum

ITIE VAN HOUT

In the textile collection at the Tropenmuseum in Amsterdam, a large number of textiles originating from the Indonesian Archipelago are represented. In fact, examples from many old and new centres of the Indonesian weaving tradition are present. Also, fine examples of Javanese batik have been collected. However, one is never satisfied and when the Tropenmuseum got the opportunity to acquire an important batik collection, we acted positively. We were able to purchase 50 per cent of the Harmen Veldhuisen collection. Mr Veldhuisen studied, collected and documented about 3,000 textiles over a period of 20 years and, moreover, he has published some very informative books on batik textiles.

This collection, containing more than 3,000 fine textiles, has been divided in two. One half has entered the collection at the Tropenmuseum, the other will return to Indonesia.

We have taken the batik collection at the Tropenmuseum as a starting point and we have sought to provide as complete a picture as possible of the Javanese batik art. The goal to compile a complete collection will probably never be reached, but now we may say that the Tropenmuseum owns a batik collection in which most of the outstanding batik centres, and the most interesting variety of styles which have developed in the nineteenth and twentieth century, are represented. Now the collection consists of some 4,000 items of batik textiles. One of the policies of the Tropenmuseum is to keep the collection up to date. The first batik was collected around 1850 and we hope to go on collecting new developing styles of batik.

The goal of this contribution to the Dunia Batik conference was to provide some information on our batik collection in its new configuration. This paper was prepared in cooperation with Harmen Veldhuisen, who generously shared his knowledge with me.

The collection comprises sarongs, *kain panjang*, *iket kepala*, *dodot*, *selendang* and wallhangings. The oldest textiles date from the first half of the nineteenth century, the most recent from the 1970s.

Because of Java's geographical position the cultures on this isle have, to a certain extent, been influenced by foreign cultures. The Pasisir, the north coast of Java, was a meeting place for merchants, travellers and men of religion (Heringa and Veldhuisen, 1997, p. 21). Influenced by Chinese, Arabs, Gujaratis and later Europeans, the Pasisir culture assimilated a number of foreign cultural elements. Wax-resist textiles were imported from the Coromandel coast; these textiles were probably *Indiennes de commande*, which means that the textiles were made according to the wishes of the buyers in the archipelago. That is why it is now considered that these Indian textiles followed the sizes, the use of colours and the design of locally-made Javanese textiles. The role played by Chinese settlers from south China was of considerable importance in the development of the batik technique on the north coast of Java. Only recently it has been suggested that batik techniques originated in China, but where and when the technique appeared in Indonesia remains a matter of speculation (Heringa and Veldhuisen, 1997, pp. 31–5).

The first quarter of the nineteenth century saw the rise of an entrepreneurial way of producing batik. By the second quarter of the nineteenth century Indo-European women began to organize the commercial production of batik *tulis* in workshops. The increase of the population on Java, the growth of trade and industry, the building of the railway and the introduction of the idea of fashion led to an increase of the commercial batik industry between 1890 and 1910. World War I led to an enormous decline in batik production, because Dutch cotton was no longer imported. After the war the decline did not come to an end: many batik makers were deprived of an income. The batik industry never reached the old levels of production (ibid., pp. 39–44).

The oldest batiks are made of locally grown, hand-spun and hand-woven cotton. Later, fine white cotton was imported from India; from 1816 English and Dutch woven cotton was imported.

The first textiles are two *kain simbut* made in a Sundanese village in West Java. A handwoven, carefully prepared white cloth was ornamented with the paste made of sugar water and sticky rice. The resist was applied by means of a bamboo stick or drawn with a finger. The natural dye, red, was rubbed into the cloth.

In the past, every family used to have these *kain simbut*. They played a role during life-cycle ceremonies, were part of a bride's dowry and were also used as blankets in times of illness or during birth (Veldhuisen-Djajasoebrata, 1984, p. 51). These *kain* show rather simple archaic motifs, like a pattern derived from the swastika and rooster.

One of the oldest cloths in the collection is a handwoven *kain panjang* from Cirebon from around 1850. It is known as *kain kelengan*, a blue and white batik. This combination of colours is valued for its apotropaic properties (Heringa and Veldhuisen, 1997, p. 21).

The next example is a handwoven shoulder cloth, a *selendang* from Demak. This kind of textile was found all along the north coast of Java. The colour red is very typical for of Pasisir batik. Chinese influences are shown by the depiction of the phoenix.

A *kain panjang* from Demak is known as a *sidangan*. *Sidangan* means that the decorations on the textiles are divided into a brighter and a darker side. This cloth is a traditional north coast batik. In the Pasisir and also in East Java, several varieties of *soga* brown were produced. Nowadays these colours are associated with batik from the principalities.

The collection also contains an old Pasisir batik, a *selendang* from Indramayu. This *selendang* was made for export to Sumatra, where the open work border was added to the cloth. On the centrefield birds and sea creatures are depicted.

The last handwoven piece is a *kain panjang* from Tuban. This textile bears a flowered pattern which can be related to the flowered textiles, the *sembagi* which were imported from India into the archipelago for several centuries. Many different kinds of textiles in Indonesia show these flowered patterns (Veldhuisen, 1993, p. 20).

The batik from Cirebon is made of extremely fine cotton, imported from India. The blue and white garment, made for a special ceremony, has an extra decoration of *perada*, or gold leaf. The pattern shows birds and several small land and sea creatures. This motif symbolizes cosmological concepts of fertility and regeneration (Heringa and Veldhuisen, 1997, p. 208).

The Cirebon holdings also include a typical *kain panjang* from around 1850. The motif on deep red is a representation of a pleasure garden for a ruler (Veldhuisen-Djajasoebrata, 1984, p. 62).

A head cloth, *iket tengahan*, also originates from Cirebon. To us Westerners, the design almost resembles art nouveau style. The pattern is a variation on the *mega mendung*, the well-known raincloud motif from Cirebon.

The *kepala* (head piece) on the collection's sarong from Lasem is characteristic of the oldest type of sarong. In the lower and upper border a pattern resembling lace scallops is depicted. The imitation lace border is considered to be an Indo-European development of the traditional snake border, which is an important pattern on Peranakan sarongs. Snakes or dragons play an important role in Javanese and Chinese symbolism – they are both metaphors for abundance (Heringa and Veldhuisen, 1997, p. 62).

Typical big lotus flowers are often depicted on batik sarongs from Lasem by Chinese entrepreneurs. Early Chinese influences can also be discerned on our *kain panjang* from Semarang. A *garuda* and a horse are depicted, but not in Javanese style, which is typical for Chinese manufacturers.

When one mentions early European influences on batik, one is referring to the products of Carolina Josephina von Franquemont and Catharina Carolina van Oosterom. Von Franquemont started her first batik workshop in Surabaya in 1840. In 1845 she moved to Semarang, where in the same year Mrs van Oosterom also opened her batik workshop (Veldhuisen, 1993, p. 21).

Von Franquemont did not sign her batik. We can recognize it because of her characteristic style and use of colours. Her batiks are famous for their rich red, blue, brown and green colours.

The pattern on the *badan* (body panel) of this sarong shows a strong similarity to the multicoloured *sembagi* textiles from the Coromandel Coast in India. Characteristic of a von Franquemont batik are the gaps in the *papan*, the border next to the *kepala* (ibid., p. 58).

The collection contains a sarong with bright colours on a beautiful cream-coloured background, which were created by the preliminary oiling of the cotton cloth, which was necessary for dyeing with *mengkudu*.

The *perada* on our von Franquemont sarong was only added to a ceremonial garment, and was usually ordered by the buyer of the cloth. The motif shows the hero Panji Inu Kertapati and his beloved Tjandra Kirana (Veldhuisen-Djajasoebrata, 1984, p. 58).

The collection also possesses another batik with *wayang* figures from the workshop of Carolina van Oosterom. European motifs are easily recognized on a sarong made in the van Oosterom workshop around 1850. To be seen are, among others, a cupid and the Dutch flag. European motifs were often taken from Dutch illustrated fashion magazines.

On this van Oosterom batik we see a Javanese royal entourage; the king in his carriage out for a ride on a special occasion. In the cavalcade we see soldiers, court dignitaries dressed in *dodot*, and servants.

Before turning to the principalities, I should like to refer to a big cloth, a *kampuh* from Madura. This batik, with bold designs, was probably made for an important person. Only the aristocracy was allowed to wear bold patterns like this.

The pattern on this *kain panjang* from Yogyakarta is called *tambalan*, which refers to garments made of patchwork. Priests and Buddhist monks dressed in garments made of many patches of cloth. *Tambalan* is a pattern worn by kings as well as by clowns. This pattern also resembles the patchwork jacket worn by the Sultan of Yogyakarta (Veldhuisen-Djajasoebrata, 1984, pp. 76 and 79).

A batik from Yogyakarta in the collection shows the pattern *sidomukti*, the meaning of which refers to the prosperity of the undertaking. During the wedding ceremony in Yogyakarta, a noble bridal couple can be dressed in clothing with this pattern. Nowadays commoners also wear this pattern at their wedding ceremonies (ibid., p. 116).

Dodot are ceremonial garments for nobles. Only half of our huge *dodot* from Yogya can be photographed at one time.

This textile from Solo is meant to be made into *celana* or sleeping trousers. The motif based on two crowned *naga* forms the side pipings of the trousers.

The Gobe sisters in Yogyakarta opened workshops there around 1900. They were followers of the English arts and crafts movement which produced a design style we now call 'Nieuwe Kunst'. The collection houses a silk headscarf made in the Gobe workshops.

Jambi batik shows more similarities in layout with Indian textiles than with Javanese batik. The collection contains only one known sarong made in Jambi, Sumatra. Known as a *kudung*, it was used as a veil to cover the head and upper body of Muslim women. Typical for Jambi batik is the lozenge-shaped design, which differs in pattern from the rest of the cloth (van Roojen, 1993, p. 161).

This *selendang* from Lasem was made to be exported to Sumatra. On textiles like these, Indian motifs were depicted for Sumatran consumers.

Dua negeri means that the cloth is dyed in two different batik centres, in this case Yogyakarta and Pekalongan. The collection also contains a *kain tiga negeri* made in three different production centres. The motifs in red were applied in Lasem, the background with motifs in blue in Demak, the brown was added in Solo.

Batik is produced in Banyumas in central Java. The collection houses a *kain panjang* from Mrs Matheson who lived in Banyumas around 1910. Because of the batik trade in Bandung, batik Banyumas became very popular. Mrs Matheson was a niece of Mrs van Oosterom and she worked in her aunt's style. The colours of this batik show the typical *matheron mengkudu* red in the border, and brown, blue and black on a cream background in the *badan* (Veldhuisen, 1993, p. 124). Also from Banyumas is *iket kepala*, a head cloth.

The *selendang* in the collection from Surabaya is decorated in different techniques. The centre is decorated in *teritik* and *pelangi*. This ceremonial cloth was part of the costume worn by court dancers.

We now turn to batik made in Pekalongan. Mrs Fisfer was one of the first batik entrepreneurs in Pekalongan. She probably was one of the first to sign her batiks (ibid.). The holdings include a sarong made by Mrs Fisfer.

When Mrs Jans' husband died in 1885 she changed her signature from J. Jans to Widow Jans; after 1900 she changed back to J. Jans (Heringa and Veldhuisen, 1997, p. 72). The museum possesses a sarong signed by widow Jans Pekalongan.

This sarong was made after 1900 and is signed with J. Jans. Among the Indo-European and Chinese manufacturers, Mrs Jans was, as far as we know, the only Dutch batik entrepreneur in Pekalongan.

The design on a sarong made in Indo-Arabian workshop in Pekalongan shows resemblances to Indian double *ikat* textiles, *patola*. The pattern is executed with the *nitik canting*. Because of the Islamic ban on depicting human beings and animals, we seldom see naturalistic images on batiks made by descendants of Indo-Arabian people. There is in the collection a sarong from Pekalongan, with a beautiful red *badan* and peacocks as fertility symbols in the *kepala*. This kind of sarong was worn by Indo-Arabian girls on the evening before their wedding.

I have mentioned many *kain panjangs* and sarongs, but batiks with other uses were also made, such as wallhangings. Two of those in the collection were made by Miss A. Wollweber and Miss S. Haighton.

One of the most important manufacturers in Pekalongan was Mrs L. Metzelaar. She was the trendsetter in renewing and changing patterns in the *kepala* and in the *pinggir*. One of the characteristics of her batik is the motif in the *pinggir*, which consists of seven leaves with three or four stylized flowers in between (Heringa and Veldhuisen, 1997, p. 79).

The collection also contains a batik signed by E. Van Zuylen. She is the best known Indo-European batik entrepreneur. She worked in Pekalongan from around 1890 until the end of World War II (Veldhuisen, 1993, p. 94).

In Pacitan on the south coast of Java the Coenraad sisters settled around 1880. They came from Surakarta and produced their batiks in the colours of central Java, indigo blue and *soga* brown. In east Java these batiks were very much appreciated as wedding presents among the Javanese nobility and well-off Chinese (ibid., p. 121).

Oey Soe Tjoen and his wife Nettie Kwee were well-known Chinese entrepreneurs who had a workshop in Kedungwuni. One of the earliest Oey Soe Tjoen sarongs from 1935 is housed in the collection.

Another one in the collection was made after the Second World War. This sarong shows the typical small dots creating shades in the colours of the flowers. This style is still made by the son and the daughter-in-law of Oey Soe Tjoen.

Shortly before the war a special group of batik was made in Kudus and Solo. These textiles feature a bouquet against a very crowded background. They were known as *buketan Semarangan*. They were sold to well-to-do Peranakan women in Semarang (Heringa and Veldhuisen, 1997, p. 44).

Also in the collection are two sarongs made during the Japanese occupation, the *Jawa Hokakai* .

The late twentieth century holdings include two garments made by Ibu Sud in 1969 in Jakarta: a garment called Malaysian dress and a shirt. Shirts like these, but in a long-sleeved fashion, were a new creation at that time.

We have reached the end of this overview. The last textiles I should like to refer to are works of batik art made by Dutch artists, such as a batik on silk by Chris Lebeau depicting four apes.

This section holds a wallhanging depicting two butterflies made in Yogyakarta. It is signed L. Baumgarten and E. Leraschi, the names of designer and the lady who made the batik. There is also a batik on silk was made by J. De Wilde, depicting two swans floating on a dark blue pond.

References

Heringa, R. and Veldhuisen, H. (1996), *Fabrics of Enchantment*, Los Angeles, Los Angeles County Museum of Art.

Heringa, R. and Veldhuisen, H. (1997), *The World of the Pasisir*, Los Angeles, Los Angeles County Museum of Art.

Van Roojen, P. (1993), *Batik Desain.*

Veldhuisen, H.C. (1993), *Batik Belanda 1840–1940: Dutch Influence in Batik from Java History and Stories*, Jakarta, Gaya Favorit Press.

Veldhuisen-Djajasoebrata, A. (1984), *Bloemen van het Heelal*, Amsterdam, A.W. Sijthoff.

PART 5
VIRTUAL AND SOUVENIR BATIKS

24 Extraterrestrial Inspiration – a Remarkable Batik from the Textile Museum Collection

MATTIEBELLE GITTINGER

In looking at a spectrum of batik patterns one can discern a continuum that stretches between the antipodes of entirely Javanese-inspired designs and those certainly deriving from non-Javanese sources. The middle ground that joins these is broad, with imprecise edges and much room for debate. In the past 20 years, serious scholars, many of whom are presented in this conference audience, have looked and found the serious and profound foundations that structure many of the traditional batik expression (Heringa, 1989 and 1993; Veldhuisen-Djajasoebrata, 1984; Tirta, 1996). These reflect concepts from a realm of thought originating beyond the individual batiker and, when given form, deny flagrant individual interpretations.

In this paper, I have gone to the opposite end of the continuum, to a remarkable batik in the Textile Museum collection that represents the extreme of non-Javanese design sources. Because we can identify precisely its source, an American comic strip published in the 1930s, we can 'see' much more of the batik artist at work and some of the conditions and mores of her time and place. (The batik 1985.51.3 was a gift of Katharine Z. Creane.)

The cloth has a *kain panjang* format worked on a very fine cotton foundation measuring 239 x 106 cm. With the exception of the black, which may be a chemical dye, the dyes appear to be the natural blue (*Indigofera tinctoria*) and brown (*Pelthophorium ferrugineum*) commonly associated with the batik work of the Javanese central principalities. The craftsmanship is of the highest level with minute filling of the background with small curls (*ukel*) and central forms with a multitude of small patterns (*isen-isen*). These are all details that fit comfortably in a Javanese scene; only the subject matter stands unique.

The figures represented come from the comic strip known as *Flash Gordon* (see Raymond, 1971 and 1990). Both in artistic style and subject matter this

227

comic strip was a startling innovation when it appeared in 1934. The artist, Alex Raymond, used daring renderings of the human body, unusual vantage points and dramatic shading. All of this matched the action-filled story line that enthralled a grateful audience. This was the introduction to the golden age of comics.

A brief look at the beginnings of this entertainment form helps to explain the impact of this particular strip. The precursors of the comic strip pictorial genre can be traced primarily in European sources up to the end of the nineteenth century. However, around 1880 a number of American daily newspapers began publishing on Sundays, and often these included a colour supplement. High points of the supplements were coloured cartoons which, over time, developed the defining characteristics of the comic form which were the sequential narrative, continuing characters and enclosed dialogue (Horn, 1976, p. 11).

However, the drawing of these early strips, as well as the subject matter, was fairly predictable. The cartoonists' humour turned on slapstick and buffoonery, although the choice of subject matter could give pause for consideration, i.e. the escaped lunatic facing an even more insane outside world; the innocent set upon by a savage society; or the little man trampled down by cruel fate (ibid., p. 7). These strips were drawn with simple lines, largely lacked shading and were often cast with a naïve feel. Throughout most of the early history of comics, they were seen as a form of communication, without much though to the art involved.

The period at the beginning of the century was a time of great explorers and the spirit of their escapades changed the comic genre by the advent of the adventure strip. Now the buffoonery and, eventually, the simple flowing line, disappeared and comics became a hybrid of their past, together with magazine illustration and the movies. The first to make the pictorial break into this new form was Raymond, who, on 7 January 1934, published the first *Flash Gordon* strip. From the time of its first appearance, *Flash Gordon* met with success, quickly surpassing *Buck Rogers* as the preeminent science fiction strip, primarily through Raymond's artistry which came to influence all subsequent adventure strips. It was translated and published in most European countries. In time, the strip was rendered in radio (1930s and 1940s), book form (1936), movies (1936, 1938 and 1940) and television (1953–54).[1]

The pictures were the captivating element of the strip, certainly not the dialogue, which was contrived, stilted and just plain 'clunky', as one writer described it (Goulart, 1975, p. 64). This historian of the comics further judged

It's probable that most of Flash's followers didn't pay much attention to what was lettered under the pictures. It was the pictures themselves – vast tableaus of lovely women and heroic men in fantasy palaces, scenes of lush monster-ridden jungles, and all that larger-than-life bravura action, all those adolescent dreams of romance and adventure so patiently given life – which seduced the readers (ibid., pp. 64–5).

The beginning of the strip in 1934 announced the coming end of the world. A new planet was rushing on a collision course towards Earth and only the work of the scientist Dr Hans Zarkov could save the world. The strip then introduced Flash Gordon, a Yale graduate and world-renowned polo player, and Dale Arden, both passengers on an aeroplane which is hit by a meteor. They bail out and eventually end up in Dr Zarkov's back yard and, subsequently, in his rocket ship which is destined to collide with, and thus deflect, the Earthbound planet. This leads to their landing on the planet Mongo and all the subsequent adventures take place in this extraterrestrial setting. It is a planet of numerous environments and creatures who could dine comfortably with Steven Spielberg. Rocket ships, ray guns, television and many machines that prefigured today's world were common ingredients of the action.

All episodes turn on the them of good vs evil. Evil is personified by Ming the Merciless, who is trying to gain control of the entire planet. Aligned against him and his forces are the elements of good – Flash Gordon, Dr Zarkov, Dale Arden and their helpers. Good always prevails in the end, although Ming is never destroyed. This basic story plot would certainly have seemed familiar to a Javanese audience, given similar themes in their own literature.

From this enormous box of potential motifs it is of interest what the batik artist selected for her cloth, what she did not and what changes and interpretations she made. Certainly there are the lead characters: Flash Gordon, Ming the Merciless. Ming is not too changed from his original rendering, but it is curious that this is the only figure not rendered in its complete form; the batik artist limits the image to the head and shoulders. Flash Gordon, who appears in many of the early episodes clad only in swimming trunks or shorts, is now completely patterned over his entire body, giving the effect of a close fitting shirt and tights. It is evident that while Alex Raymond sought to portray a vigorous, muscular action hero, the mores of the batik artist dictated more circumspection.

Exotic humanoid forms of the planet Mongo obviously delighted the batik artist. She adapts with great veracity the winged bird men, the spined lizard

Illustration 24.1a   Detail of a batik from Central Java depicting scenes from a *Flash Gordon* comic strip (Textile Museum; gift of Katharine Z. Creane)

Illustration 24.1b   Detail of a batik from Central Java depicting scenes from a *Flash Gordon* comic strip (Textile Museum; gift of Katharine Z. Creane)

Illustration 24.1c **Detail of a batik from Central Java depicting scenes from a *Flash Gordon* comic strip (Textile Museum; gift of Katharine Z. Creane)**

men and the horned villains that challenge the hero in various weekly episodes. The fantastic creations of Mongo are captured with particular zest. Dragon-inspired forms, prehistoric birds, sea creatures and exotic avian forms vie for space with time machines, rocket ships and guns of battle.

In the comic strip, the episodes occur within the threatening jungles and seas of the planet Mongo. The Javanese artist tames these environments into a benign floral landscape familiar to the batik world.

The batik artist has obviously relished the exotic angles from which the comic strip figures were frequently rendered. The turning of a body, the rendering of a back view, a body in flight seen from above, are all perspectives beyond Javanese artistic tradition, yet the batik artist captures these with apparent ease and fidelity and incorporates them into her cloth.

What she does not use is probably even more revealing of the batik artist's time and culture. Although the cloth surface teems with life and the movement of many forms, there is no interaction among the forms, nor do we sense the violent action of the strips. Individual characters, machines and animals are faithfully recorded after the original drawings, but not their interaction. Her cloth is not narrative. Even more remarkable to the external observer is that while the pages of *Flash Gordon* are strewn with alluring ladies, no women appear in the batik rendering. It may have been that the ladies were too alluring for the Javanese time and place, but banished they were.

The batik artist has not just borrowed individual design elements. Rather, she has created a lyrical planet Mongo with its cast of heroes, villains and demons, each endowed with unique filling forms of scales, small flowers, dots, circles and dashes in various combinations. The figures appear on the cloth surface as they are needed to make the composition writhe and turn and all the remaining areas are filled with the swaying floral and plant forms. It is truly a batik Mongo in the finest sense. Alex Raymond would be envious.

How did these comics come to the Javanese batiker? One can only speculate. Certainly many Dutch publications reached Java in the 1930s and we know *Flash Gordon* was translated into German, Italian, French and Spanish and, presumably, Dutch, although King Features syndicate, which furnished comic strips to the European syndicator, Bulls Presstjant, has no records extending back to that time. A majority of the characters on the batik derive from the original strips numbered 1–81, which appeared between 7 January 1934 and 11 August 1935 and the remainder from strips published from 1936–38. They may not have been sent to Java as part of newspapers, but in a comic book format, which had already evolved in the form we know today by 1933. Also, the strips were reprinted many times, so the 1930s need not be the date of this cloth.

However, if certain stylistic characteristics are considered in comparison with other batiks believed to have been worked in the 1930s, there is a strong suggestion that the cloth was made in this period. Several batiks have the faces of a majority of the forms, both human and animal, rendered in an opaque black probably of chemical origin. One of these, with a pattern of battleships and soldiers, believed to have been made in Pekalongan c. 1930 by unknown hands, depicts rows of black-faced soldiers and swimming black-faced fish (Tirta, 1996, Plate 22). Similar black-headed fish appear on a batik attributed to Haji Ehsan from Pekalongan that is dated c. 1930–40 (Heringa and Veldhuisen, 1996, p. 146). In all other regards, however, these batiks bear no stylistic relationship to the finely worked Flash Gordon batik. A textile that does approach this cloth in its level of fine craftsmanship was given to the Textile Museum by K.R.T. Hardjonogoro in 1979 (1979.6.10). It was made in Surakarta in the 1940s and has numerous birds with dark black heads (Gittinger, 1979, p. 32). This particular idiosyncrasy of utilizing black heads coloured by an opaque chemical dye on batiked figures seems to have begun in the 1930s and continued into the Hokokai period of the 1940s. It is a feature of few batiks after this. Earlier batiks may have depicted people with darkened faces (Heringa, 1996, p. 54), but not the intense black of these forms. Given these considerations, the Flash Gordon batik was probably made in the latter part of the 1930s or early 1940s. Because the cloth reflects the work of a truly skilled batiker, it is hoped that that person might be known to some in this audience of batik experts. It would be a fitting closure to be able to place a Javanese name next to those of Alex Raymond and Flash Gordon in the records of the Textile Museum.

Note

1 In comments following the presentation, Ibu Ratmini Soedjatmoko reported that she and her brother read *Flash Gordon* in *d'Orient* in the late 1930s while living in Banjumas. The magazine was one of several that came in the *lees trommel*, a circulating rental packet that supplied foreign reading material. *Flash Gordon* was extremely popular and widely known throughout the upper classes, according to Ibu Ratmini.

References

Gittinger, M. (1979), 'Conversations with a Batik Master', *The Textile Museum Journal*, Vol. 18, pp. 25–32, Washington, DC.

Goulart, R. (1975), *The Adventurous Decade*, New Rochelle, New York, Arlington House Publishers.

Heringa, R. (1989), 'Dye Process and Life Sequence' in M. Gittinger (ed.), *To Speak with Cloth*, pp. 107–30, Los Angeles, Museum of Cultural History.

Heringa, R. (1993), 'Tilling the Cloth and Weaving the Land. Textiles, Land and Regeneration in an East-Javanese Area', *Weaving Patterns of Life* (Indonesian Textile Symposium, 1991), pp. 155–76, Basel, Museum of Ethnography.

Heringa, R. (1996), 'Batik Pasisir as Mestizo Costume' in *Fabrics of Enchantment*, pp. 46–69, Los Angeles, Los Angeles County Museum of Art.

Heringa, R. and Veldhuisen, H. (1996), 'Emblems of Colonial Power' in *Fabrics of Enchantment*, pp. 139–47, Los Angeles, Los Angeles County Museum of Art.

Horn, M. (1976), *The World Encyclopedia of Comics*, New York, Chelsea House Publishers.

Raymond, A. (1971), 'Into the Water World of Mongo' in *Flash Gordon*, New York, Nostalgia Press (originally published by King Features, 12 April 1936–30 October 1938).

Raymond, A. (1990), 'Mongo, the Planet of Doom' in *Flash Gordon*, Vol. I, Princeton and Wisconsin, Kitchen Sink Press (originally published by King Features, 7 January 1934–4 August 1935).

Tirta, I. (1996), *Batik. A Play of Light and Shades* (Vol. 1) and *A Collection of Batik Patterns and Designs* (Vol. II), Jakarta, Gaya Favorit Press.

Veldhuisen-Djajasoebrata, A. (1984), *Bloemen van het Heelal*, Amsterdam, A.W. Sijthoff.

25 The Challenge for Batik in the Year 2020: Art, Commodity and Technology

AMRI YAHYA

Introduction

It has been widely realized that technology has a wide spectrum impact comprising value systems, culture and conceptual changes. Consequently, the government tries to carry out programmes of reculturization and reconstruction. The reculturization programme refers to the readiness to face the advent of the new era (millennium) with a new culture, namely the technological and industrial culture. This will give an impetus to the rise of new problems, including: first, a shift from a traditional value system to a modern one; second, a culture of technological working ethos; third, a social structure; and fourth, a means of living. Essentially, these need a basic change (restructuralization) necessarily made by the Indonesian government as well as by Asian-Pacific countries. One example worth presenting is batik.

Batik, an art product with specific values characteristic of Indonesian society, needs special treatment in its development. One possibility is to use batik as a specifically Indonesian commodity within tourism. The emergent problems are, among others: the prospect of development according to the international quality standards; the context of philosophical values; the development of clothing and as a medium of expression; and the establishment of batik institutions in the context of maintenance. For these reasons, this paper tries to present a case study concerning batik in the context of tourism.

Batik and Tourism in the Global Perspective

Globalization is an appropriate expression regarding the growth of the international economy in the 1980s and 1990s. It has a rapid, significant, strong and sometimes radical influence on present conditions. It breaks through every nation's economy, both in developed countries and in developing countries.

For developing countries, especially Indonesia, facing such globalization is not a matter of accepting or refusing, but of making positive use of the maximum profit and minimum loss by reducing its disadvantages (Spillane, 1994, p. 88).

One perspective is that the greater the world's economy is, the stronger the smallest players are (Naisbitt, 1994, p. 4). This means that the government should take decisions on policy seriously. Some craftsmen leave their foster companies, try to break into the free market and make an illegal connection with the buyers at the expense of quality control. As a result Indonesian Standard Quality is not achieved, and this degrades the image of Indonesian products.

The Loss of Pearl

There are several types of tourism, but the most outstanding one is service tourism (see Budihardjo, 1995). To attract tourists, it is necessary to develop modes of attraction, such as souvenirs. This may give new inspiration to souvenir vendors. It is realized that batik is only a small element of the factors supporting success in tourism. From another perspective, however, batik is associated with the identity of Yogyakarta in particular and of Indonesia in general. This is a means of promotion with its shift to an unintended commercial vision. However, in order to secure sales, the price of batik may be reduced and this may result in unfair competition, which in turn degrades the image of batik as a central attraction for tourists.

Tourism may be viewed from two sides: firstly, it can turn a retailer into a wholesaler, which means the improvement of the craftsmen's economic life; secondly, the spirit of entrepreneurship rises. Conversely, there is a disguised degradation in ideas, concepts and economy for the sake of profit. This may be right from the perspective of living standard improvement, but may diminish the quality of batik development by the standards of Indonesia's ancestors. Batik is composed of a high philosophical content, meaningful motifs and appropriate materials for clothing. But nowadays, for the sake of economic satisfaction, batik is sold through stratified packages; a direct contact between a buyer and a retailer is unavoidable. As a result, Malaysia can defend batik as her intellectual property right in the world.

For the time being it can be concluded that a nation's intellect plays an important role in national development. A wide vision is needed to avoid undercutting and the attendant risk of lower standards. One way to achieve such a success is returning to the initial objective: tourism promotion. Batik may be regarded as a string of pearls with high value.

Towards Symbiosis Amid the Wave of Global Culture

The process of cultural transformation is indispensable after the effects of globalization on Indonesia and other Asian-Pacific countries. Cultural contact gives a new meaning to the assimilation process of tradition and modernity as well as the attendant legal changes. Likes and dislikes in this tough exchange are common. Traditional patterns may shift when modern culture is adhered to by the young generation; modern culture may replace the glorious, old tradition, which still serves as the backbone of the nation's existence. There may be some differences between the expectation of the young generation and that of the proponents of the traditional patterns. This may result in a crossing of cultures.

In a similar way, the condition of batik undergoes the same problem. Batik in its historical context has a long history with symbolic meanings. Batik depicts complex life, for example the *semen* motif (Javanese *semi-an*, meaning seedbed). This motif depicts growth and life and signifies fertility. From a technical perspective, the motif shows the transfer of learning and values. It is an attempt to hand down the skill of making batik characterized by accuracy, patience and care. The motif gives a meaning to life: improve your life as plants grow in the universe.

It is necessary to note that life's struggle demands financial aid and spirit. Batik exists in tradition, but now has to struggle in the modern era to preserve its value. It is necessary to adapt to achieve glory; there may be a sacrifice for it. Originally batik existed meaningfully and now it is tied up in the unity of form and economy.

Concerning form, batik now is immersed in various visual media such as: a medium of expression (for example, painting); a medium of prestige (for example, a party gown made from silk); and many others. There is also a shift from the handmade methods to painting and printing techniques. At a glance, printing gives a derogatory connotation although socially it does not. For those who can afford it, the possession of the original handmade technique applied on silk confers prestige. But for those who cannot afford such luxuries,

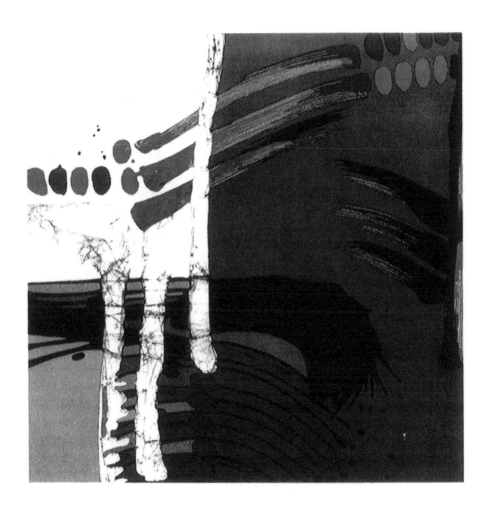

Illustration 25.1   'Hanya Tinggal Arang': batik on cotton, 80 x 80 cm

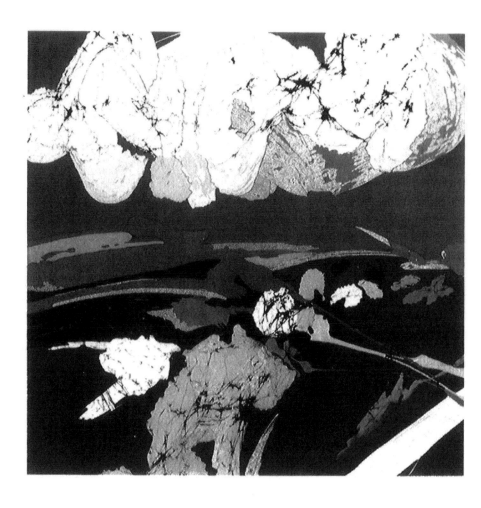

Illustration 25.2   'Lebak Merah': batik on cotton, 75 x 75 cm

Illustration 25.3 **'Daun-daun': batik on cotton, 80 x 80 cm**

Illustration 25.4   **'Rumput di Hutan Terbakar': batik on cotton, 80 x 80 cm**

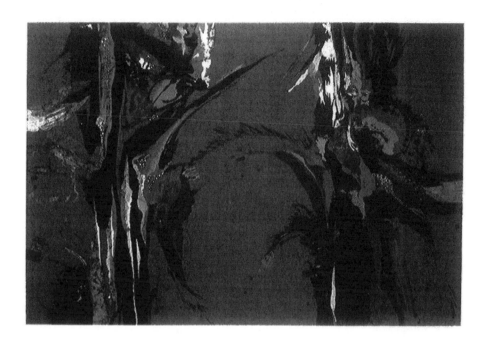

Illustration 25.5 'Hutanku': batik on cotton, 300 x 200 cm

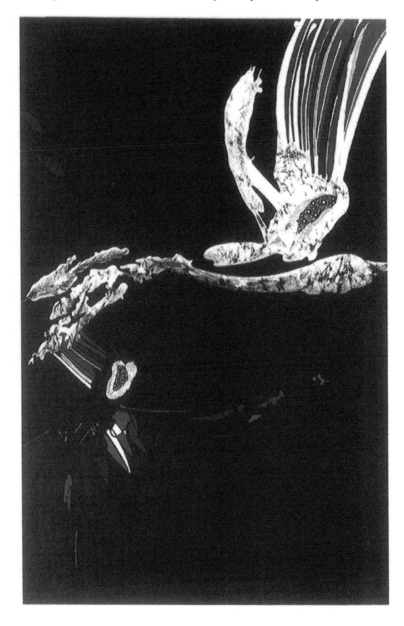

Illustration 25.6 'Hutanku Tinggal Arang': batik on cotton, 120 x 80 cm

Illustration 25.7 '**Pohon-pohon Arang (O, Hutanku)**': **batik on cotton, 300 x 200 cm**

printed batik is adequate to satisfy their needs.

Economic unity means that batik is adaptable (for example, batik for clothing). Originally worn in Javanese clothing as *jejarik*, batik is now made into dresses, skirts, bed covers, etc. Quality may decline due to the replacement of the medium with various materials. People, however, have to accept and take the risk of the shift from symbolic meaning to economic meaning in accordance with consumers' tastes.

Oddly enough, almost everyone admits, though informally, that batik still exists in the life of Indonesian people. For example, at a wedding party, a funeral ceremony, a celebration day and a formal meeting, people still wear batik as an identity marker.

Therefore it is the duty of modern generations to formulate a symbiosis with the modern condition in order not to perceive batik as marginal to culture. The symbiosis with modern culture through reactualization can be realized in various forms such as: a batik-wearing movement, batik uniforms, a batik year, etc. It is necessary to establish a collaborative attempt at development in Indonesia.

In addition, the global world is characterized by a pluralist view, including an idea of exploiting batik as part of life. Batik can be a commercial commodity (for a trader), a dress (for a designer and a consumer), and a medium of expression (in painting).

From the concept of pluralism, we still appreciate batik as a work which will develop its form, motif, medium and application. The existence of printed batik, which was initially debatable, is not a serious problem in the 1990s. People have to know the rationale for the way in which they look for an identity through clothing. The 'have-nots' will choose printed batik with a low-quality medium. But there is still hope for the losers in batik that they should maintain the quality of form, technique and material in order not to degrade the overall quality.

Institutions of Development

The Diplomatic Paradigm

To realize the ideas proposed above, it is necessary to look back at the existence and position of batik in the discourse of institutionalized development. Neither the government's nor private attempts have been effective in protecting batik. The legal aspect has not been supportive from the industrial perspective.

As stated above, the deconstruction of people's economic patterns gives a chance for a retailer to make direct contact with a buyer. The government may lose income through lower taxes. A syndication of buyers and craftsmen making direct contact may avoid government tax. The government, however, may play a protective role and help to guarantee the quality of craftsmen's work.

The role of diplomacy will give an increasing demand if the two aspects run in harmony. National perspectives and business should be unified for the sake of the nation's development.

Batik, which has become the people's spirit, is legitimized through the formal events held by the government and followed by other groups of people. Sociologically, Indonesian society is paternalistic. Therefore, an example of wearing batik at a formal occasion will always be followed. On the other hand, the development of batik in the context of expression needs space for exhibitions so that its development in Indonesia will achieve its potential.

Promotion Board

The diplomatic paradigm is proposed as government policy and it needs to be guided and promoted through a well-prepared plan. It is necessary to establish an institution dealing with preservation (for example, Centre of Batik Research and Bamboo Craft) in cooperation with universities in order to conduct research and develop a sustainable plan. Private attempts to improve batik industries in villages are supposed to lead to the satisfaction of the quality standard of ISO 9000.

The Promotion Board will be able to provide a bridge between businessmen, technocrats, researchers, craftsmen, sellers and exporters to achieve a unified whole (interdepartmental integration). Thus, integral development can be achieved.

Formal Development

So far the development activity in batik carried out by the government has not been parallel with formal development. The formal development in schools is planned in stages, at elementary schools, at junior high schools, and at senior high schools, as well as training packages for beginners. This pattern will continue to be developed, alongside the marketing plan, and in cooperation with the Batik Promotion Board.

Non-formal/Informal Development

The non-formal attempt through direct guidance to the craftsman has become nearly extinct because of its partial nature. The craftsmen depend on the buyers and the price is determined by the buyers. This attempt should be supported by batik associations under the Batik Promotion Board in an attempt to: first, conduct a national research study on batik and the impact of its development; second, build a national museum to provide a historical description and to develop ideas and development patterns; third, write books and research reports on batik; fourth, hold a competition on batik article writing to give an appreciation of batik at the initial stage; fifth, hold a batik motif design competition, which has not been carried out because most activities are batik exhibitions; and sixth, hold a workshop on training in research into colours, materials and motifs to explore new possibilities for batik.

Closing Remarks

Finally, our faith in batik depends on us. Will we treat batik as an orphan or a treasured child? Batik is a pearl growing naturally from generation to generation as part of tradition, functioning mostly as *jejarit*. Due to the situation, batik has developed as art and commodity. From the art perspective, we can make use of batik as a painting technique, and an idea of motif development, etc. As a commodity, batik has contributed much to people's lives in the form of dresses, gowns, napkins, curtains, chair cushions, etc.

The spread of batik as a commodity and its symbiosis with tourism should maintain its quality. Similarly, the batik village is no longer humble, but can serve as a centre of integrated development. The economic deconstruction is not a competition for the sake of personal profit without considering the future. Economic awareness and the improvement of academic quality will accelerate quality. Concerning the acceleration of batik as a commodity, it is hoped that this chapter will result in some ideas to institutionalize batik development in Yogyakarta. The Batik Development Board and the Batik Promotion Board can cooperate with government institutions to provide mutual benefits.

References

Boediardjo (1991), *Pariwisata dan Kebudayaan di Indonisia (makalah)*, Jakarta, Konggres Kebudayaan.

Boediardjo (1995), *Pariwisata Budaya (makalah)*, Jakarta, Konggres Kebudayaan.

Hadiwarno, B.H. and Wibisono, S. (1996), *Memasuki Pasar Internastional dengan ISO 9000*, Ghalia Indonesia, Jakarta, Sustem Manajemen Mutu.

Kleiden, I. (1995), *Pergeseran Nilai-nilai Moral*, Jakarta, Perkembangan Kesenian dan Perubahan Sosial.

Neisbits, J. and Aburdene, P. (1994), *Megantrend 2000* (terjemahan: Budianto), Jakarta, Binarupa Aksara.

O'Hara-Deveraux, M. and Johansen, R. (1994), *Global Paradox, Menjembatani Jarak, Budaya dan Waktu* (terjemahan: Agus Maulana, Linda Saputra), Jakarta, Binarupa Aksara.

Spillan, J.J. (1994), *Pariwisata Indonesia*, Yogyakarta, Lembaga Studi Realino-Kanisius.

Susanto, S. (1990), *Seni Kerajinan Batik Indonesia, Balai Penelitian Batik dan Kerajinan*, Yogyakarta, Lembaga Penelitian dan Pendidikan Industri, Departemen Perindustrian RI.

26 Lifestyles in the Borderless World: Marketing Sarawak Textiles as Cultural Identity Products

MOHAMMAD ZULKIFLI

This paper will rediscover Nusantara textiles and clothing from a historio-graphic and economic point of view. Nusantara is known in the Malay world as maritime Southeast Asia, with a special focal point on Malaysia and Sarawak. In this paper, one could see words like 'borderless' and 'globalization' mentioned frequently and how these concepts of trading in the 1990s is linked with intra-Asia trading before European imperialism in the fifteenth century.

The chapter will address 'globalization' in the discussion, with a little history on textile trading in Southeast Asia until the present market. The chapter will briefly discuss the world-famous *pua kumbu* of Iban, the famous *songket* of Malay and other ethnic motifs of Sarawak, and how these motifs became the source and inspiration for contemporary textile designers in Malaysia.

Then the chapter will measure the marketing environment trends and styles in textile and clothing industry. The discussion on Sarawak and Borneo in terms of their uniqueness of lifestyles, multi-ethnic character, rare flora and fauna and rare breeds of animals. These will be the unique features in the product-printed and hand-painted batik of Sarawak. In parallel to that, I will also be discussing cultural identity and multi-ethnicity.

The paper will give a brief background of how and why hand-painted batik arrived in Sarawak. Hand-painted batik in Sarawak is basically a new product and needs to be developed more. Few companies have discovered the technique, and both the product and the strategy for local markets need to be more adventurous and creative. So I will refer to some examples of batik based on Sarawak ethnic motifs. What I am interested in is to develop new products and new creations looking at Sarawak's wealth of arts, culture and nature without having to sacrifice the indigenous arts. New arts could be developed by looking at ready inspiration such as: the life of local people, the hornbills, the Rajah Brooke butterfly, crocodiles, orang utans and ferns, orchids, raflesia etc.

Eco-tourism, cultural and nature tourism will be discussed with reference to the new Sarawak government policy and promotional campaign known as 'CAN' (culture, adventure and nature). The discussion of cultural identity, globalization and cultural products will take place alongside Sarawak batik as new product. The chapter will also include a discussion of trends and market needs and show how the product could find a marketing niche in the global business era.

Market segmentation, market targeting and niche marketing will be discussed in relation to new products. The chapter will discuss a creative arts marketing approach and products for the tourist specialized market: eco-tourists, new affluent local market, yuppies and baby boomers, batik as wearable art, fashion and clothing, decoration and painting, and so on.

My interest in textiles and clothing trade started to mature when my mother brought me to Singapore when I was nine, in the early 1970s. We travelled all the way from Kota Bharu, Kelantan (northeast Peninsula, bordering Thailand) and stayed with our relatives in Bussorah Street, off Arab Street, for at least a week. We went shopping for Indian *palikat* and batik *jawa sarong* in Arab Street; Chinese and Javanese *palin* and floral cotton, silk and polyester cloths in Arab Street, *geylang and tanjong katong* bazaars and shops; English and Javanese, wool and polyester pants materials in High Street; and of course C.K. Tang shirts in Orchard Road. We brought back the goods to Kelantan to be sold. This was my normal routine every time we came to Singapore during my school holidays from the age of nine to 16 years old.

Later in life, after going to business school, having studied marketing and international marketing (unfortunately never becoming a textile merchant), I travelled to various parts of the world. I collected a few pieces of textile – old and new – which made me wonder and want to know more about the textile trade, especially in the Southeast Asia region.

Kelantase women, for example, are known as entrepreneurs, since the seventeenth century when Raja Cik Siti Wan Kembang was the ruler (the first female ruler in Kelantan). They brought back to Kelantan, textiles from every corner of the world: Korea, China, Thailand, Indonesia, India, and others. In Kelantan itself, textiles were produced and exported to other parts of Malaysia and abroad. *Kelantan* batik and *songket* have their trademark and style, changing every month according to current fashions and trends. But they have a style of their own.

Sarawak, another state in Malaysia, is famous for *pua kumbu*. I first encountered *pua kumbu*, which had been brought to Kuala Lumpur's central market, after it was converted into a craft market in 1986. I met a Sarawakian entrepreneur designer, Fatimah Abang Saufi in her boutique called *Layang-Layang*. Later, a few other entrepreneurs discover the route to supplies of *pua kumbu*. But of course Fatimah did not stop there but went further to design knitted wear based on *pua kumbu*, designed specially for foreign markets.

In 1990, I was presented a printed *pua kumbu* on cotton, designed by a Sarawak-based company, Fabrico. I only achieved my first possession of *pua kumbu*, woven by a colleague's mother, when I got posted to Sarawak in 1993. Probably that was the first point of my journey of collecting ethnic textiles and wanting to know more about them.

A Little Background of Textiles in Nusantra

From a historical perspective, Nusantara was a centre of trade and cloth production before the rise of Sri Wijaya, Majapahit and Malacca. According to Maznah Mohammad (1995), the textile industry in the Nusantara could be divided into four phases. The first phase involved the Indian Ocean linking Southeast Asia to India, the Middle East and ultimately Rome. The second phase involved intra-Asia trade with the Chinese and strengthening political relations. The third phase was referred to as European imperialism, where the English East India and the Dutch India Companies manipulated the Nusantara routes, products and markets. The fourth phase concerned the Malay textile industry in the nineteenth century when each community developed their own industry and market. In the mid-nineteenth century, textile imports into the Malay Peninsula dwindled. Local hand-loom production was thus stimulated during the transitional period catering for the local market which was experiencing a dearth of foreign textiles. Though by no means comparable to India, textiles were produced in greater bulk than in earlier years in order to fill regional market demand.

It was a few years before textiles from the power looms of Europe superseded hand-loom textiles, whether Malay or Indian. This last significant phase determined the conditions under which the hand-loom industry here began to acquire its present form. Remnants of these old Malay weaving centres still survive, hinting at a varied and fascinating past.

Nicholas Bernard, in his book *Living With Decorative Textiles*, talks about international trade with the region being dominated by exports of luxury commodities such as spices , rare and scented woods and dyestuffs (especially indigo) for over 300 years from the fifteenth century onwards. Textiles from mainland Asia were used by Portuguese and Arab coastal traders as goods for barter, so bringing to the islands weaving designs and influences from afar, especially the Indian subcontinent.

M. Gittinger, in *Textiles and Tradition in Indonesia*, refers to loom-decorated textiles, *ikat* work, that delighted the Dutch settlers and such cloths were exported to Holland over 100 ago, to make striking bedspreads, curtains and loose furnishing cloth. Certainly, the *ikat* textiles and textiles woven on simple backstrap looms demonstrate the continuation and development of a weaving tradition that has ancient and complex roots.

The resist-dyeing techniques of *bandhani* and *bandha* are known elsewhere by their more familiar titles of tie and dye or *pelangi* and *ikat* work respectively. *Bandhana* textiles are prized as veil cloths, saris, shoulder cloths and turbans. The tie and dye work is found throughout India, especially in Gujerat, Rajasthan, Orissa and Andha Pradesh. It is the monopoly of Hindu and Muslim communities of professional weavers, dyers and printers. These textiles were among the major commodities traded with Southeast Asia. These techniques could have inspired the islanders who traded with the Indians and created their own motifs, symbols, designs and images.

In Java, Sumatera and Bali the *patola* cloths have been imported from at least the fifteenth century onwards, and used for religious and court ceremonies. Historically a luxury commodity from Gujerat, antique examples are treasured as festive clothing and shrine decorations amongst the traditionalist upper class families of the region.

By contrast the professional weavers and dyers of Orissa and Andhra Pradesh have since late nineteenth century and more recently, adopted and adapted bandha techniques most successfully. They are flourishing by means of cooperative systems of production and marketing. Textiles are woven and patterned to suit the ephemeral tastes in colour and design of the market at home and abroad (Lynton, 1995).

Bark cloth was widely used throughout Sarawak when cloth was scarce. For instance, during the Japanese Occupation of World War II, when practically no new cloth was imported, many villagers brought their bark beater out of the rafters, where the humble instrument had been gathering dust, and revived the craft. Not only clothing, but also blankets and even mosquito nets were made of this stiff material. Before this emergency, some indigenous people never did much weaving and generally relied on trade with their neighbours for good clothing.

Bark cloth was also employed for *pua* and clothing, particularly jackets. Designs were also painted on or applied by stencil. The Kenyah people in particular stencilled exotic representations of human figures and a dragon-like creature called the *also*, combining dog elements. These were favourite subjects in media other than bark cloth, such as beaded and shell appliqué.

Lynton points out that appliqué work is generally associated with village embroidery tradition and is also found in Bengal and Orissa. The attachment of pieces of cut cloth and other decorative objects to a plain ground in an appliqué techniques and the joining together of layers of cloth with running stitches to make a quilted fabric are both ancient methods of ornamentation. Gujerat is justifiably famous for appliqué work of covers, hangings, trappings and household decorations. These could have been another source of inspiration for the people of Nusantara.

Away from the coastal enclaves of the Chinese and Malays, the Iban peoples traditionally pursue a lifestyle of shifting cultivation centred on their longhouse settlements. Their most famous textiles are the *pua* warp-*ikat*ed blankets and hangings reserved for ceremonial use. Woven in the shade and cover of the longhouse gallery the *ikat*s are decorated, often on an attractive brick-red ground, with electric geometric patterning that almost overwhelms the senses. Only from a distance may one observe that these designs link to form interlocking chains of animal and human figures.

The Iban comprise nearly one-third of the total population of Sarawak, of about two million. They live throughout the lowland and coastal areas of Sarawak, residing mostly in the longhouses along rivers. From their genealogies and legends, the Iban claimed to have moved, looking for better land, from southern Borneo, into the territories which are now Sarawak, around the middle of sixteenth century.

Textiles are central to the traditional Iban of life. The *pua kumbu* is a popular native handicraft of the Iban; translated literally, *pua* means blanket, *kumbu* means warp. Together the two words means 'grand blanket'. However, the *pua kumbu* is seldom used as sleeping blanket, but has a more spiritual meaning.

For centuries these Bornean people have practised shifting agriculture, primarily to cultivate rice, cotton, and occasionally dye stuffs. *Gawai* are connected with agriculture, warfare and the festival of the dead. In the olden days, the *pua kumbu* was very much an integral part of day-to-day affairs and special rituals of Iban society. One or more pieces of *pua* were hung prominently in the midst of joyous gatherings. Decorated textiles appear only on these occasions, and their use has the same ritual and symbolic complexity as

Illustration 26.1   Iban *ikat* weaving

the ceremonies themselves. These are occasions for great merrymaking, when finely dressed young people fraternize and elders flaunt their wealth. But the backdrop for the these activities is solemn ritual in which decorative textiles are used not for display, but as integral parts of a sacred pattern.

The *pua* may also form a small room or enclosure on the gallery, built to contain the dead and their mourners or a new mother and her child. Sacred textiles were also used in the rituals associated with headhunting.

The art of making a *pua kumbu* is a much-valued tradition passed on from mothers to daughters. The scores of designs and patterns incorporated into each *pua kumbu* are reflection of the beliefs and values of the Iban people. The Iban also draw inspiration for their *pua* patterns from animals, which play an important part of their daily lives. The most common animals found in *pua* are crocodiles, snakes, birds and lizards.

The *pua kumbu* is a creation of a mythological and religious heritage of centuries, representing a woven picture of history. *Pua* invites unhurried viewing so as to study the images of legends, religion, history and autobiography, woven into intricate patterns. They also believe that certain designs are granted to them in their dreams. The line between this world and the other is indistinctly drawn for a creative artist.

In spite of the incursions of the modern world, textiles and decorative arts still persist in Borneo. Although the subject matter may now include vivid colours and modern, often quite amusing, forms, the style remains distinctive.

Gawai Dayak, after joining Malaysia in 1962, became a common official holiday for all the Dayak-Iban, Bidayuh and Orang Ulu, in recognition of the existence of Dayaks in Malaysia. This festival is probably the biggest contemporary festival celebrated in which *pua kumbu* are used in ritual offerings to the gods on Gawai Eve. During this time, all relatives are back from wherever they are, wearing their best dress, serving the best food and decorating the house to receive guests from near or far. There are a few other different *gawai* such as *gawai antu* and *gawai batu*, where they use the *pua kumbu*.

Sarawak Malay *songket*, another known textile of Sarawak, with distinctive motifs, differs from Peninsular Malaysia (Kelantan and Trengganu), Brunei or Kalimantan Malay *songket*. *Songket* is a rich brocade of one colour, silver or gold, or mostly gold thread. The piece is of uniform length, mostly, patterned with a sprinkling of isolated stars, flowers or loose geometrical design over the main part. The intricate panel in the centre or at one end is known as the head of the cloth. Animal and human designs are not allowed in *songket* weaving in accordance with Islamic prohibitions. One of the differences of Sarawak *songket* from Kelantan, Trengganu and Brunei *songket* would probably be the curvy motifs adapted and adopted from Dayak design. The Iban too have their own *songket* but they preferred it to be called *sungkit*, where they used gold, silver and multicoloured bright threads. These may have been adopted and adapted from the *songket* of the Malays. Iban *sungkit* has only floral and geometrical designs.

Textiles from other ethnic groups include beaded work, appliqué and patchwork. The Iban, arguably Borneo's most skilled weavers, used bark cloth for making either stout war coats and rough wears or for padding woven or beaded garments. Other groups like Bidayuh and Orang Ulu also used these materials for the same purpose.

Amongst many contemporary Iban weavers there is a tendency to over-simplify designs as well as methods. *Pua kumbu* of hand-spun cotton-dyed with natural vegetable dyes have given way to synthetic cotton yarns and modern colours brewed from cheap instant aniline dyes. This dilution in method is probably due to the constraints of time and an unwillingness to use the laborious and meticulous *kebat* method of old. But it also betrays a disturbing lack of interest demonstrated by contemporary weavers, of the esoteric knowledge and rituals of traditional *pua kumbu*.

Souvenir-type textiles are produced in bulk at crafts centres and associated villages. These joint ventures are sponsored by concerned agencies in the hope of preserving traditional handicrafts, sustaining ethnic entrepreneurship and marketing indigenous art to a wider audience.

The innovative production of contemporary quality Iban silk textiles employing traditional methods, has generated interest within the fashion industry, both local and abroad. Commercial frontiers are being explored for Iban textiles. It is uncertain and questionable whether such ventures will be able to stimulate and sustain the pure, sacred tradition of weaving, as the two methods are in contradiction. The contemporary method is to mass produce ethnic textiles for a large tourist market where commercial viability is the driving force. The traditional method is the sacred expression of Iban values demonstrating status and spiritual maturity, with no thought of profit margins.

The future for Iban traditional weaving, however, does not look too bleak. It is not unrealistic to envisage that the mass restoration of traditional Iban textiles weaving as a sacred spiritual undertaking amongst Iban women can be realized through rediscovery of Iban culture as a whole. This can be achieved through a true understanding and appreciation of Iban traditional values and lifestyle by the Iban themselves. There is an urgent need for the Iban to appreciate the tradition and values so deeply held by their forefathers, lest this rich culture is lost.

Nicholas Bernard (1989) in his investigation of contemporary tribal weaving communities, showed that, on the whole, the art of traditional weaving is surviving and developing. Indeed, it is reassuring to find that, in the minds of our new world of instant communications, rapid travel and consumer ephemera, the ancient tradition of tapestry weaving is still flourishing among many indigenous peoples.

Today, these weavers are entering a new era of rapid and unprecedented change in the development of their vibrant and energetic folk art. Inspired by market forces and by all kinds of cultural influences – philanthropic collectors, peace corps volunteers, local merchants and tourists – the weavers are adapting, copying and replicating old work and inventing new styles and compositions.

It is impossible to determine whether any of these directions should be actively discouraged or encouraged. So long as the melting pot of creative and technical influences continues to be stirred, so much the better. It is enough that the looms are still being worked by hand, producing textiles to give both pleasure and use to someone, somewhere.

Now, we could probably see *pua kumbu* with old traditional designs as well as contemporary designs intermingle in the market. But most of them are probably using new materials, colour and threads. Some commissioned works by certain individuals or organizations could revive old designs. The bottom line is the supply of raw materials and demand of the product from the market.

Sarawak contemporary textiles can be categorized into the following varieties:

- *pua kumbu* for jackets, sarongs or short skirts woven by Iban women in their longhouses for family use;
- *pua kumbu* woven by Iban weavers for designers and institutions using old motifs but excluding sacred motifs or spiritual attachment;
- traditional-style *pua kumbu* made for institutions and private commercial organizations as souvenir items for tourist market. The designs are normally of non-spiritual motifs, while the size also been reduced to a side table cover size;
- new *pua kumbu* using simplified contemporary motifs, not necessarily in Iban or Sarawak style often combining new colour and techniques. Mostly made by new generation artists, trained by art schools either locally or abroad or in federal government craft centres;
- Sarawak hand-painted batik produced by artists from national and regional government agencies, commercial designers or individual artists. They

Illustration 26.2   First production of printed Sarawak motif on cotton

Illustration 26.3   Printed Sarawak motif on cotton rayon in various colours

Illustration 26.4   Sarawak motif in batik

Illustration 26.5   Printed Sarawak motif umbrella

use Sarawak ethnic motifs with batik technique on silk, cotton, and rayon, etc. These are becoming popular, especially among Malays and government officials. This is influenced by fashionable batik in Kelantan, Trengganu and Kuala Lumpur, and the Islamic movement in Malaysia as well as international design houses such as Versace;

- as far as Sarawak is concerned, there is only one company, Fabrico, producing mass-produced printed textiles on silk, cotton, linen, rayon, crepe, etc. These textiles are not necessarily produced in Malaysia: most of them are produced in Thailand, Singapore, Bali and Kalimantan, Indonesia, Philippines, Korea, Japan and China. Most of the motifs used are either from a single ethnic source or a mixture or inspired by internationally known designers like Christian Dior, Gucci and Versace. Designs from Japan, Korea or China are popular.

The market for these categories of textiles could easily be divided into the following segments: the mass market approach especially for a cheap range of textiles, mainly cotton, polyester and rayon; and the high end approach for more expensive materials like woven silk, printed silk and linen.

Again these two segments could be divided into three types of market: the local market (daily clothing and utility like curtains, upholstery, bedsheets and so on); the specialized market (artists/textile collectors buying old *pua kumbu*); and the tourist market (local and foreign). Some of the latter buy these textiles as a means of supporting the community as a sustainable development activity.

References

Adorno, Theodor W. (1991), *The Culture Industry*, London, Routledge.
Bernard, Nicholas (1989), *Living With Decorative Textiles*, London, Thames & Hudson.
Buma, Michael (1987), *Iban Custom and Traditions*, Kuching, Borneo Publications.
Gittinger, M. (1985), *Textiles and Tradition in Indonesia*, Oxford University Press.
Hill, O'Sullivan and O'Sullivan (1995), *Creative Arts Marketing*, London, Butterworth and Heinemann.
Jenks, Chris (1993), *Culture*, London, Routledge.
Kotler and Amstrong (1996), *Fundamentals of Marketing*, New York, Prentice Hall.
Lynton, Linda (1995), *The Sari: Styles, Patterns, History and Technique*, London, Thames & Hudson.

Mohammad, Maznah (1995), 'The origins of Weaving Centres In The Malay Peninsula', *JMBRAS*, Vol. 68.

Ohmae, Keniche (1990), *The Borderless World*, London, Fontana.

Ong, Edric (1986), *Pua: Iban Weavings of Sarawak*, Sarawak, Sarawak Atelier Society.

Ong, Edric (1996), *Sarawak Style*, Singapore, Times Editions.

Sarawak Atelier Society (1987), *Sarawak Legacy*.

Siti Zainon, Ismail (1994), *Textile, Tenunan Melayu*, Kuala Lumpur, DBP.

Welck, Gisela V. and Welck, Kavin (1985), 'Indonesia textiles', Symposium, Rautenstrauch, Josest Museum, Cologne 1991.

27 Batik Route: Batik and Tourism

FRED W. VAN OSS

For over 40 years, since the age of 16, I have been working with textiles. For the longest period I was a shirt manufacturer, with a final production of 10,000 high quality shirts a day. Later, I worked as a consultant for the textile and clothing industry and in 1986 I was appointed as the direction of the Textile Museum in The Netherlands. But in between I studied at university during the evenings. And so, at the age of 46 years, I graduated as a lawyer specializing in social-economic and business affairs. During the period I worked as a consultant, I also worked for a national institute for education where I assisted in making educational programmes for textiles schools. It is strange, but during the period that I worked in order to try to explain to other people what textiles mean, it was the period that I had the feeling that I learnt the most myself.

This experience resulted in some books, some video films and, during the period I acted as the director of the Textile Museum, in exhibitions as well.

Other Cultures

For some of these exhibitions I went to other countries, such as the USA (1991–92), where I assembled an exhibition about the weaving of the Hopi and Navajo Indians. I was very much impressed by the quality of the weaving, but even more by the culture of the people I met and by the landscape of the southwest. Here you will find famous areas such as the Grand Canyon, the Petrified Forest, Mesa Verde, Oak Creek Canyon, and Canyon de Chelly.

So I decided, as well as creating the exhibition and the catalogue, to organize cultural tours to this area. It was a great success, not only for the people who made the trip, but also for the local Indians who were able to sell a lot of their own products at really good prices.

Because of this, I also had a lot of contacts with representatives of the Council of Europe in Strasbourg, France. The aim of the Council of Europe (established in 1949) is:

> to achieve greater unity between its members to safeguard the European heritage and to facilitate their economic and social progress through discussion and common action in economic, social, cultural, educational, scientific, legal and administrative matters and in the maintenance of human rights and fundamental freedoms.

In fact I have seen this organization mainly as a kind of human rights organization. But they are very active in stimulating tourism as a way to get better contacts and understanding between different people from all over Europe. It was a great surprise for me that these professionals used textile especially as an item in order to encourage people to travel. And so they developed 'silk roads', 'wool roads' and 'linen roads'.

From that moment on I developed a kind of professional interest in 'cultural tourism'. What is 'tourism' in fact, and what is 'cultural tourism'? Are there any definitions? The World Tourism Organisation defined 'tourism' in 1993 as follows:

> The activities of persons during their travel and stay in a place outside their usual place of residence, for a continuous period of less than one year, for leisure, business or other purposes.

In Europe there exists an organization named ATLAS (European Association for Tourism and Leisure Education). They have in mind starting a similar organization in Asia, and this will probably start in 1999. ATLAS also has a definition for cultural tourism:

> The movement of persons to cultural attractions away from their normal place of residence, with the intention to gather new information and experiences to satisfy their cultural needs.

'Cultural tourism' is beginning to be recognized as an important economic force in Europe, and one hears jokes such as 'cultural tourism is the oil industry of France' or the 'General Motors of Italy'.

What is Cultural Tourism?

When you start to think about tourism, it is easy to think that this is a very new activity, but this is not the case. For many hundreds of years, people have been travelling, mainly as a form of study or education, but also for pleasure. It was not unusual for travellers to visit factories in other countries as well as cultural monuments.

But the 'big boom' of cultural tourism started after the 1960s, mainly with German tourists who possibly had the most money.

You have to realize that culture is a very large field of human interest. In 1989 a list was published that contained, amongst other points: archaeological sites, museums, architecture, music, dance, theatre, festivals, complete cultures, subcultures and arts and crafts.

My Batik Experience

In October 1994 the governments of Indonesia and The Netherlands agreed a cultural treaty. One of the points mentioned in that agreement was that staff in the Museum Tekstil in Jakarta would like to get more experience. Luckily enough, the city of Tilburg in The Netherlands houses The Netherlands Textile Museum. This museum generally is considered to be the largest and most complete textile museum in the world. As a coincidence I used to be the managing director of this museum, so I had the task of 'doing something'.

I thought it probably would not be good to act as a kind of teacher. We decided to do a joint project and, while working together, we could show how *we* used to work. In this way our colleagues in Jakarta could pick up useful ideas and translate these to their own Indonesian way of thinking and working.

We decided to make an entirely complete batik project, which means an exhibition, a book and a video film. But, with perhaps more that 1,000 exhibitions, more than 100 books and more than 10 films made before by others, why do the same? So we decided to be different – different in all possible ways.

Most exhibitions, books and films concentrated on Yogyakarta, Surakarta and Pekalongan. But after some studies we found that there were at least 42 cities and villages that made batik. We decided to visit all these areas. We wanted to be very professional, so we were accompanied by our own photographer and our own designer. While sitting at my computer, with all kinds of maps of Java, I tried to find the best possible way to travel in order to visit all the places. Finally I found the way to do it. It would be a 2,000 km trip. I used 'WordPerfect' and after finishing my text and route I had to store it and to give it a name consisting of eight plus one, two or three characters: the 'Batik route' (BATIKROU.TE) was born.

We prepared the exhibition, the book and the film. The exhibition could be seen in Tilburg and some parts of it in Jakarta as well. But after dismantling an exhibition, it is lost, of course. But the book is still available. It is written in three languages: Indonesian, of course, English and Dutch. The Indonesian title is *Batik Sukma Jawa* or *Batik, the Soul of Java* in English.

The book explains the history of batik, shows how batik is made and how it can be used, covering 12 regions. We only displayed 66 pieces of batik, but in a way we thought it was very clear to show the basic ideas and the most important differences between the regions. For us it was also very important to show human beings and also the relationship to nature. Maybe the 12 pictures that were used as an introduction for the regions can illustrate this intention. The original pictures were taken by our photographer, Frans van Ameijde. The film is made as a 30 minute VHS-PAL video. It is available now in a combination of English/Dutch, but I still hope it will be possible to make the 'Bahasa Indonesia' edition as well.

Batik Tour

Naning has a special place in my heart. During the year of the project, I visited Indonesia five times and the village where Naning lives three times. She was very proud to show the batik of her village. But a few months later I returned and gave her a very large copy of her picture. She was at once the hero of her village. A few months later I returned again, with 16 tourists who bought a lot of batik in her very poor village. And I gave her the catalogue with her picture on p. 159.

In October 1996 we also undertook a three-week tour of the batik areas with 16 tourists. Most of these tourists had a particular interest in textiles, but in any case they were very interested in culture. So I can say these were real cultural tourists.

We arrived in Jakarta and went to Bogor, Bandung, Garut, Tasikmalaya, Ciamis, Cirebon, Trusmi, Tegal, Pekalongan, Kedungwuni, Yogyakarta, Solo, Demak, Kudus, Juwana, Rembang, Lasem, Kerek, Tuban, Ponorogo, Pacitan, Tulungagung, Blitar, Kediri, Mojokerto, Surabaya, Madura, Gresik and many other interesting places as well. Unfortunately we were not able to visit Indramayu and the regions of Banyumas and Purwokerto.

We visited all kinds of batik factories and batik ateliers and artists. I can mention some names: Harjonagoro, Iwan Tirta, Oey Soe Tjoen (Ibu Yenni and Ibu Indrawati), Ibu Masina, Moch Sjoleh, Purnumo, Moch Rusli, Sendari,

Skrikandi and of course some in Yogyakarta such as Amri Yahya, Ibu Winotosastro, Ibu Pranoto Wiyarjo.

Each traveller paid about US$2,000 for the trip (rate at September 1997). The price of the plane ticket was less than US$1,000, so more than US$1,000 per person was spent on hotels, meals and travelling inside Indonesia. Besides that, there were the so-called 'personal costs'.

For a country it is very important that a high percentage of the money really does come to the country itself. During other conferences I have heard a lot of complaints about the fact that some countries, which attract millions of tourists, do not earn any money. All the profit seems to go to the big international tour operators which are usually housed in the countries of origin.

But most interesting for you, of course, is the amount of money that has been spent while visiting the batik factories and batik shops. I do not have an exact accounting of all the people, of course, but I am sure that at least 200 original pieces of batik were bought. This quantity can be converted into money, but also, and in fact more importantly, it can be converted into months (in fact, years) of labour.

A reasonable batik requires 1–3 months of work. So if we accept an average, I am not far off if I say 30 years of work. This means that each of the people in my group were buying 2 years of work. It may be that that group showed more interest than average groups of textile tourists will have. You should not forget that nearly everywhere we were treated really very specially, with good food and even special performances with gamelan and dances. But even if others would buy only half the quantity, it still means that one good tourist will be equal to one year of work for the batik makers.

The Future

Imagine that every day, or nearly every day, say 300 times a year, a group of 16 people arrived in Indonesia in order to make that same trip: it would give 300 x 16 = work for about 5,000 people. But we should not just look at the results for the batik makers themselves but also for their villages and cities, for hotels, restaurants and all other kinds of people who can earn money by giving services to tourists. There are a lot of publications available specially on this subject. There are no good reasons to go into detail now, but generally speaking, I can tell you that the effects would be incredibly great.

But we have to take care. We should *not* try to bring more and more tourists to the same villages and batik makers where my group has been.

Otherwise, the positive effects will be spoiled. Research has to be done in order to find *all* the batik makers and *all* the interesting places in their surroundings. This will be better for the local people themselves and for the tourists as well. Also, better knowledge about hotels and restaurants is essential, otherwise there will be another kind of problem. But after this research many more tourists than the 5,000 I mentioned can come, see and buy.

Earlier I mentioned the danger that all profits would leave the country. For that reason I think that 'batik tourism' should be handled with care. It may be that textile organizations all over the world could cooperate with textile organizations in Indonesia in organizing the tours. Tour leaders with a good textile background are essential.

There is one other important point of view that might be very positive. I have been talking about the real, handmade batik. The people organizing batik tours will mainly be interested in genuine batik. But I am sure that the fact that so many people will take pieces of batik all over the world will have an influence on fashion in the world. That means that there will be a growing market for machine-made designs that have been inspired by original batiks. And I am also sure that these produces, printed in Indonesia and assembled into all kinds of clothes, will be more appreciated because they are made in Indonesia. So this side-effect will mean employment for a lot of people and will be a considerable support for the Indonesian economy as well.

28 The Threads That Tie Textiles to Tourism

KAYE CRIPPEN

Batik is popular internationally in the late 1990s, with Kenzo showing Indonesian batik sarongs and velvet batik in his recent collections. With this international focus on batik, now is an excellent time to get increased worldwide exposure for Indonesian designers and artisans, and secure more international retail placements for high-end batik products. Since fashion cycles can be short, it is important to act quickly; there is a narrow window of opportunity to introduce batik to a variety of new consumers.

The fashion trend for batik could also influence tourist demand for batik. Tourists also shop at airports and hotel gift shops. So where are the batiks at the airport shops in Indonesia? Batik Keris has such shops. So the first recommendation is to get more visibility for batik at airport, hotel shops, and museum shops.

The debate on what is batik continues. It means different things in Indonesia, Singapore, Malaysia, Thailand and China. Purchasers, including tourists, may have different understandings. I have given up trying to tell friends that the shirt they bought is not *batik tulis*; they don't want to hear it. My advice is that the consumer will decide what batik is and they control the purse strings. The best you can do is to try to educate those interested. However education is expensive and time consuming. Simple communications are best but they must also be engaging. Many consumers tell me they find understanding batik overwhelming. You must demystify it, while keeping it intriguing.

The threads that tye textiles to tourism are getting stronger, with many tourists choosing a destination simply because of a textile technique such as batik or *ikat*. With a growing international interest in traditional textile techniques, greater press exposure via specialized publications and newsletters, an increase in *textourism*, or specialized tours, featuring textiles as a sole or key element. Remember that tourists can be domestic tourists, regional tourists, or international tourists. Tourists are not just tourists, but members of different

271

communities. One Laotian weaver, who had lived in France, worked out her colour preference system for what tourists from a variety of countries preferred.

Tourism and Traditional Textile Linkages

Impact of Tourism on Textiles

Railroads to Native Americans Tourism has had a profound impact on many traditional textiles ranging from helping to maintain the tradition to assisting in the alteration of the art or craft. In the American southwest, for example, Native Americans rapidly altered the motifs on their *Indian blankets* to suit their perception of tourists' demands. They made smaller sizes and incorporated tourist icons. Today, one can buy a model of a woman in traditional costume seated in front of her upright wooden frame loom weaving a small rug.

Luckily the art form survived tourism and many levels of these blankets or rugs are now available for the tourist or serious collector, from the fine old antique blankets, to the new high quality pieces which command premium selling pricing. Lesser quality new and older items are also available. All coexist in the market. The annual arts fair in Santa Fe is a time when tourists come to see and in some cases purchase the best of show. Prizes are awarded for best of show from any category. This serves as a good model for the batik industry.

Molas in San Blas Molas is a type of reverse appliqué, done by the *Cuna* Indians of the San Blas islands off Panama, which was changed to include new motifs which were popular with some tourists. The key changes which greatly reduced the time required to make a *mola* was the reduction in the number of layers used and in the size. This also results in a reduction in the number of colours appearing in the final form and makes it less visually interesting, but still suitable for many tourists. The original function switched and became simply a tourist souvenir. Last year, the Textile Museum in Washington, DC offered small circular *mola* patches in the Textile Museum gift shop.

Even in the early 1970s in California, retailers such as Pier 1 and others were scouring Mexico and other areas for textile items to market along with other types of traditional craft items such as basketry. Textiles are very decorative and make great rugs, tablecloths, bedspreads, napkins, place mats, pillows, draperies, wallhangings and dresses. Although this type of retail format gave broader international exposure to hand-woven textiles, in many cases the higher volumes needed combined with the lower price points, had a major impact on the type of hand-woven traditional textiles produced. It changed the art form and hopefully was economically rewarding to the weavers. Decisions were made as to how to make the textiles more quickly in order to meet these price points. In Mexico and some other areas, cooperatives run by women weavers are popular outlets for tourists.

In recent years, an increased demand for the higher and middle market traditional textiles has also occurred. The interest in textiles has increased as has the number of popular publications, associations, study groups and tours which help the lay person to understand traditional textiles. This understanding has helped fuel demand.

Change is inevitable but it is also necessary for the survival of many textile traditions. My main wish is that some weavers will be able to continue their traditional textile weaving using traditional techniques, motifs, and colorations. My second hope is that the high end will continue including the use of traditional motifs, as well as new interpretations. Reproductions must be labelled as such, but it can be difficult to prevent them from being sold if some unscrupulous person decides to age them.

The second goal is for more women to be able to produce traditional textile products as a viable way for many to make a living. Many areas are changing from the barter system to using money. *Bisnis Indonesia* noted that many women from Lamongan in *Jawa Timur* are giving up their traditional weaving to come to Jakarta to be *bakso* sellers or going to work outside the country. Hopefully your traditional textile arts can be preserved and one way is through tourism, though more development projects are needed.

Artisans must understand consumer needs and make products that will sell, and have market access. In addition, they need money to finance their supplies and to live on while they make many time consuming textiles. In many cases, development projects give training to artisans with the goal of making them self-sufficient. One such case was reported in the *Jakarta Post* where Oxfam personnel were helping Balinese wood carvers understand the market for carved cat, *kucing*, products. These carvers were getting 50 per cent of total sale price deposit before carving started. The cat craze should be capitalized upon in the batik market as well. An artist in Jamaica is selling batik paintings of pineapples via the Internet.

In Jambi, I was told that the batik pattern I wanted was for men. They did not understand that in my circle of friends no-one would know; and if they did, it would not matter. This is where cross-cultural product development specialists can work with local artisans to help them understand product development opportunities.

There is a growing market for alternative trade products which companies like Oxfam from the UK and other markets in numerous countries, including Canada, Japan and the USA. Oxfam markets these products mainly through mail order catalogues. Others are marketing via the Internet. Alternative trade is a speciality niche business. Oxfam's definition of fair trade is exclusion of exploitation of child labour or women, use of environmentally friendly products and healthy working conditions. Women's wages is an international issue worth monitoring.

Many *Hmong* who came to the USA sold their reverse appliqué in order to survive. Most quickly learned to modify their products to meet market demand. One innovative lady made stuffed Christmas tree ornaments; she found a product that people wanted.

The Ching Mai Hill Tribes Research Center has tried to encourage young women that weaving can offer an alternative career choice to the sex trade. One researcher reported that it was a losing battle. Many view their mother's traditional weaving skills as not being modern, and they are desperately seeking to be modern at very high social costs. A key problem here is making a product the tourist wants at the right price point. The main item produced by one group did not attract much tourist interest.

There have been a number of training programmes in Thailand including the one for *mat mee*, silk *ikat* weaving, in northeastern Thailand under Queen Siriskrit's sponsorship. Gift shops are prevalent in the airport featuring Thai *ikat* textiles, apparel and accessory items. A number of non-governmental agencies and individuals such as Percy Vatsaloo have also helped with weaving related projects.

In nearby Laos, many have been instrumental in reviving their weaving industry including the Lao Women's Union. Kantong won the UNESCO prize and has received some international publicity. Carol Cassidy, an American UNDP worker who is a weaver, has been able to focus much international publicity on her hand-weaving workshop; she assists Laotian weavers in recreating traditional patterns as well as innovating new ones. Carol Cassidy has had numerous exhibitions and sales in international settings. Major international newspapers as well as specialized publications have written about these weavings. One expert on Laos mentioned that she had probably got more press for Laos than anyone else.

She trains Laotians, who in some cases have no previous weaving experience, and says that she can tell whether they can become a weaver when she puts them at the loom. This is the advantage of being a weaver herself. Many other programmes were not good at identifying people's potential and interest in becoming good weavers. In addition, Ms Cassidy uses computer colour matching technology to recreate the old colours using synthetic dyes. One product that was well received were long shawls with golden motifs woven in the ground; she now makes smaller ones. She has received custom upholstery orders. In addition, they have been developing a small-scale programme to raise silkworms as a crop substitute.

Mary Connors, author of *Lao Textiles and Traditions*, has documented Laotian textiles. She leads textile tours to Laos for the Hong Kong Textiles Society. She knows the weavers, the villages and has extensive knowledge of the motifs and textile traditions of Laos. Many other weavers are active in producing textiles for the domestic and tourist market. Tourism has helped stimulate demand. One village specializes in producing items for expatriates, including traditional textiles for weddings.

The above examples led me to wonder whether *textourism* can be an alternative to other types of tourism popular in Thailand and Indochina. *Textoursim* is increasing: tour groups are becoming more specialized. Tourists with special interests do not want to just see the usual sites. Tourists want to see *batik, shibori,* double *ikat* production, etc. The Textile Museum in Washington, DC and the Craft and Folk Art Museum have been offering tours for a number of years. In Singapore, the Friends of the Museum organize many tours. One such was a textile tour of northeast Thailand. Another example of a tour which would be difficult to do alone was the tour to visit minority areas in China led by Gina Corrigan. Due to a well-planned itinerary and a seasoned guide, this group saw a large number of villages with diverse textile traditions. This would be almost impossible for most to do on their own due to the rugged terrain, lack of transportation as well as the language barrier and a lack of contacts. This group had many people who were very familiar with textiles; at the end of the day there was a debriefing session where they discussed what they saw. This type of group interaction also makes this type of group travel more rewarding. Different people have different perspectives so the educational element can be enhanced.

In Alaska, the musk oxen are herded and the fibre from their underbelly, *quivit,* is the finest fibre in the world. A development project wanted to give Native Americans in the region a chance to earn money from this exotic fibre. However, their traditional garments were made from animal skins so chewing to soften skins and stitching through tough skins were skills they knew. The project taught these Native Americans how to spin and knit. The fibre was actually better suited to knitted items. They made gloves, scarves and other items for tourists and export.

Australian aborigines, known for their traditional artwork, got even more international exposure when their designs on T-shirts became an international success. An interesting project which featured cross-cultural exchange with aboriginal artists visiting Yogyakarta to learn the batik process resulted in an exhibition. The batik images were alluring and were nicely adapted to batik. This type of exchange should continue. Indonesian artisans can also get new insights to designs this way.

Japan has a long textile tradition and has a wealth of experiences for the textile tourist. Kawashima Textile Museum and factory, offers many classes. In 1996, an advanced *Shibori* workshop was offered. Another unique hands-on experience was at Ken-Ichi Utsuki's *Aizenkobo* workshop where one could purchase a white handkerchief or scarf, already tied, and do the dying themselves with the help of the owner. Japan, also has many traditional festivals, such as the autumn festivals where participants wear traditional costumes such as the one in Arimatsu where the men wear *shibori*. Japan also has numerous textile exhibitions.

Hong Kong has long had many textile tourists seeking to buy exotic textiles. Currently, some high quality reproductions are showing up in the market which some vendors try to sell as being older. This is always a concern with reproductions. Art markets abound in Shanghai and Beijing that sell textile items. There are many speciality museums in China including Hangzhou, Suzhou, Nanjing. The Suzhou Silk Museum has a new museum plus an outstanding collection of older-style looms and textile techniques with artisans demonstrating them. The museum has realized an outstanding profit due to good business management of the operation. They have a beautiful traditional style dining area and a gift shop with a wide variety of well designed textile products including reproduction items from their workshops such as silk velvet burnout shawls. They are in a good location and have a number of tourists visit their museum.

Jen Wen at the Nanjing Brocade Research Institute won the Gold Cup in China for artistic merit for recreation of an ancient emperor's robe. This institute trains students in the ancient art of weaving on the draw loom. The institute also makes reproductions for needs within PROC and for other museum outside China. However, it has been difficult for them to make money for this operation. They are seeking to develop more items that appeal to tourists. Jen Wen noted that tourists take photographs beside the souvenirs that were intended for them to buy. He did also made silk leno gauze for an Italian designer.

In Malaysia, the government sponsors weaving training programmes; these have undergone some recent consolidation. *Kraftagan* offers craft items for sale at their outlets. Innovative university programmes such as the one at UNIMAS where Narzalina Shaari and Mohamad Zulkifli bring their creativity and design background to the students in a progressive programme.

Singapore is notable now for its new direction in museums displaying textile items and its sale of older textiles through dealers and the informal sector. The increase in expatriates has helped drive demand for textiles as it has in Hong Kong and for textile related tourism there. Study groups through the Friends of the Museum often are the focus of study of traditional textiles.

Sudha Patel works with women in India in the *Kutsch* to produce wearable art that has strong appeal to western women, as well as Indian expatriates in the region. Her garments are sold in India and in the region through informal sales. She mentioned that they work carefully with women, who have low self-esteem, to gradually improve the quality of their products without directly criticizing their production.

Indonesia has a large number of weaving and traditional textile sites that are still active. Beautiful scenery, interesting culture, good food, and numerous festivals makes for a winning tourism combination. Let's hope that the tourist industry can assist in keeping those textile traditions alive by encouraging traditional weaving practices not only for the tourist but also for one's own pride and cultural continuation. Women living close to tourist areas often have a better chance of earning a higher income than those in more remote areas. A woman weaving when tourists come into her shop may have a better chance of selling the tourist the item. Some do prefer to buy from the weaver. Many tourists can tell the authentic weaver from the 'weaver on display sitting at a loom going through the motions'. Women working in the batik industry face different problems from many other women who produce traditional textiles since they do not generally market their own products directly to the consumer.

The author arrived in Tenganan in 1985 the last year that older women were weaving, and has witnessed the decline and revival of their weaving tradition. Tenganan, in east Bali, is the only village in Indonesia that produces double *ikat* textiles. Double *ikat* production is very complex and is only produced in three countries of the world. The people of this village are referred to as the *Bali Aga*, the original Balinese, who inhabited the island before the influx of Majapahit settlers and refugees from East Java. The villagers main responsibility is to keep the village pure; they observe many ritual days and duties and adhere to strict rules.

During the mid-1980s, the last of the older weavers retired. A few weavers remained and at one time only one of the women from the young unmarried women's association was learning the technique. Traditionally, the young women learned the tedious technique through these associations. Urs Ramseyer, an associate of Alfred Bühler, has documented the long process on video. He and others encouraged the women of the village to continue, concerned that their long weaving tradition might not continue. However, there was little interest and weaving in the village appeared to be in decline. The colours produced were not deep. Only the small sized breast wrappers were made with the simplest motif, the *sanan empeg*, which did not have any curvilinear forms and was therefore easier to produce. Few tourists came to the village; many thought that weaving in the village was in decline.

However, due to increased tourism and sales of *geringsing* more women are weaving and there are a number of younger students now. Tourism has done what textile experts could not do; it has motivated more women to learn and competition has encouraged them to improve their weaving. Although tourism has changed the village in many ways, the renewed interest in weaving is a positive one. The author has identified many ways for the villagers to continue improving their technical weaving, dyeing skills as well as their marketing skills

Tenganan has a greater number of tourists attending their annual *Samba Usamba*, a ritual village purification. The occasion involves a re-enactment of the ancient *pandanus* leaf wars, with unmarried women swinging on the old wooden swings. The *kain gerinsing* cloths may be seen in their cultural context.

Forming New Innovative Linkages

Possible linkages exist between the traditional textile industry, NGOs and companies. Indonesian designers can work with traditional textile artisans. Linkages with international museums are possible in order to supply their museum shops.

Tourism associated with festivals is expected to increase; already many tourists plan a trip around a specific eveny. In summer 1997 the *Kraton* Festival in Cirebon drew domestic as well as some international tourists. The highlight was seeing the various costumes of the various royal representatives attending the festival, as well as the parade and exhibitions. The latter included one from the Museum Tekstil in Jakarta, which displayed batiks from Cirebon. There are many festivals in Indonesia so this represents a major opportunity for textile artisans to sell tourist items. Some festivals are sacred and there is a fine line between what is and is not appropriate. In such a case, keep the commercial area far from the ceremonial areas. Tourists must be advised as to appropriate behaviour and dress. But they also need more information on ceremonies, the dress and the textiles.

Innovative ideas are essential in developing tourist products. Consider offering a batik course where participants stay at a nearby bread and breakfast decorated in batik. The studio could offer various levels of batik courses. The Batik Research Institute offers a good programme but instruction is only in *Bahasa Indonesian* and requires a long time commitment. Tourists often want to make something so some courses could focus on the outcome of a product (e.g. a duvet cover, pillowcases, placemats, apparel or accessory item).

Another possibility is for the tourist to develop a design concept and work with local batik artists to make their customary patterns into fabric or even a garment or interior product. First the student would develop a concept based on a theme and either work with an artist or create his or her own drawing. Then the colorations would be developed and perfected before the artisan would make the actual yardage. It is possible that a whole small scale industry could develop around meeting the needs of tourists. Other possibilities would include: demonstrations; hands-on work; special lectures on the history of batik and current batik production in the area; tours to outlying villages; and lectures on safe use of chemicals, dyes, and disposal of waxes. Word of mouth is an important component in the tourism industry as well as Internet coverage. It is helpful to use a team to generate ideas and test them out on various target tourists.

The next linkage connects to the large textile and apparel industries in Indonesia. One example of this linkage is the new line *Tirta* line designed by Iwan Tirta for Great River International (GRI), the largest Indonesian apparel producer, one of the largest in Asia. Sunjoto Tanudjaja, President and Chief Executive Officer of GRI and International Apparel Federation President in 1997, realized the role GRI could play in promoting batik to the youth of Indonesia.

Many are concerned that Indonesia, which is changing rapidly, and has a high youth population, will become a nation where the young forget their traditional textiles, fruits and foods. In Singapore, the government encourages use of batik dress for official government occasions; it is also common in Indonesia.

The government in Indonesia has launched a 'love Indonesian textile products' campaign and in Thailand the government promotes 'buy Thai'.

The difficulty is that consumers buy what they want or perceive to be fashionable. So it is not surprising that in the MTV and Hard Rock Café era, advertisements for Levi's jeans may be more appealing than a slogan saying 'buy local products'. The USA had their 'buy American' campaign, which was never really very successful because consumers decide what they want to buy.

GRI has previously produced well known international brands under license agreements. Less promotion is generally required to sell well-known brand names than to create a new brand name. The launch of the men's casual wear line was at the Model's Café in December, 1996 in Jakarta. There is also a stand-alone boutique at *Mal Anggrek* which features only the *Tirta* line. Many of the designs use touches of batik and many of the prints and T-shirt patterns show strong batik influence.

In addition to targeting the young male Indonesian, this line is also reporting good sales at some tourist areas in Bali. Marco Gandasubrata, General Merchandise Manager for Private Label & Casual Wear at GRI noted that an advertising campaign is being developed to support this line. This line has received quite a bit of publicity through editorials The costs of developing and promoting such lines are enormous. GRI is to be complemented in assisting in updating the batik concept. It takes time to refine such a line and to promote it. This latest design direction for Iwan Tirta shows the breadth of his design capabilities, including: custom and couture design, apparel and scarves, interior textiles for the home and commercial settings, wallhangings, and casual wear. Designers can link with other firms, as Iwan Tirta did with Schumacher, to develop beautiful interior textiles using batik on silk Jacquard fabric for the interiors.

Other Indonesian designers use traditional textiles as inspiration. They can provide excitement that can help create an image for the tourist and international textile and apparel markets. However, a more pro-active role must be taken in order to secure this publicity. It does not come easily. It takes planning, time, and resources. Publicity is not free; it is expensive to have someone develop the story ideas, press kits, and stay in touch with the press. However it is usually cheaper than advertising. It must be targeted at the international presses need for story ideas.

Bin House appeals to a range of women. Ah Bin says many consider her designs lighter. So just as we see the 'light' trend in food we now see it in batik, which can be used to position the product. She is also to be congratulated with her upper end pieces. She has a good understanding of who her target consumer is and how to reach them. She has retail shops in Jakarta,

Singapore and Japan.

Ardiyanto Pranata is to be commended for his wide line and his experimental work combining batik and *shibori*. His products sell in Japan and France as well as in Indonesia. He designs a wide range of products at various prices. Harry Darsono uses batik and painting on silk as inspiration even for costumes for *Julius Casaer*, as well as creative couture. He did a project with traditional weavers to produce *songket* in Sumatra which received television coverage. In addition, he trains young people in the textile arts with special needs. Just as I asked why linen had not been done, Ramli showed linen in his latest collection. Linen, one of the most comfortable of all fibres for the tropics, also has unique high wicking properties that could be incorporated into the batik as a design element. Poppy Darsono, Ghea, and Carmanita use traditional textiles and batik as inspiration. Carmanita is developing lines which are popular with young women.

There is much room for experimentation with fabrics that are in fashion including stretch fabrics, crinkled fabrics, laminated batiks, burnout batiks and many more. There is also research going on into producing a higher quality wool batik. And don't forget those easy-care blends or premium microfibres and stretch fabrics with *Lycra*. It is hoped that other large companies – either fibre, textile, or apparel manufacturers – will form linkages with talented Indonesian designers to take their message to a broader audience. I have developed various batik concepts that resemble what we did in the fibre industry to stimulate new design approaches. One must watch fashion and colour trends for new ideas but not be afraid to innovate by combining new textile base cloths with batik concepts.

Marketing Aspects

There are two interesting marketing problems. First, there is a wide variety of consumers; batik consumers are not homogenous. The second is that when we say batik, it has a wide variety of meanings. Batik illustrates the concept of micromarketing where markets for products becomes more and more specialized. Batik in itself is a speciality product, but then think of all the types and levels of products and it becomes infinitely more specialized. Some want high quality old textiles, others will buy high quality new textiles. There is another market for wearable, high quality apparel and matching accessories. There will always be a market for less expensive products.

Marketers segment the marketplace, target the customers they choose from segments they think they can serve and then position their products to fill the wants and needs of their target markets. There are many ways to segment markets. The batik market may be segmented in various ways: tourist versus non-tourist designation, age basis, socioeconomic basis, or geographic basis. Designers need to use market segmentation analysis then design products that meet their target customers needs. This can be used for designing new tourist concepts. It is difficult to be all things to all consumers, although it is possible for the same line to appeal to a diverse range of consumers. Many designers successfully market products for various consumers by changing the brand name, differentiating design and retail outlets.

The next issue is how to reach the customer. The role and importance of the Internet are powerful. I pulled up over 900 references to batik using one search engine. Some were obviously not related to the textile, but still many were offering products for sale. One artist in Malaysia who offers batik paintings via Internet had a surprise category which says 'buy all 19 for just under US$10,000'. The Internet allows one to reach niche targets.

The first task is to find them; the second is to develop the market and reach them. There are many new possible markets. For example, the ecoconscious consumer may find batik appealing since it is rather timeless and does not need to be quickly discarded.

Asia, for example, has many potential customers. At a recent conference in Singapore, I showed several relatively new pieces of batik to an audience composed of many Chinese from Singapore and PROC. Those from China in particular were unfamiliar with the batik look and were most interested in Chinese-influenced motifs such as the clouds from Cirebon and an altar piece from Pekalongan with a dragon. The Singaporean Chinese wanted to know the source.

Other possible strategies include: encouraging loyal customers to become heavier users; value-adding by making wearable art and other creative apparel; and custom design and high quality apparel made to order during the tourist visit. There is a whole market segment out there that is interested in this concept and there are publications that target them. There are many strategies and tactics to explore including labelling, describing the technique and motifs, focusing on the artists; and innovative packaging and product design.

Recommendations

First understand your potential consumers and develop products to meet their needs. Think creatively in designing new products, new motifs, and look for new ways to market products to the target customer. Follow the trends in fashion fabrics and colours. Use motifs that might appeal to the tourist who might be viewing their first *durian, rambutan* or whatever.

Get more international exposure for batik. Develop an industry and personal action plan. Indonesian designers and artisans must get more exposure in the international arena through specialized publications and newsletters that reach those interested in fashion and the traditional textile arts. Do not forget national press and the Internet. Think of batik-related stories that are newsworthy, that the press can use. Some may want to increase their linkages with large firms to develop casual wear or other types of lines such as swimwear. Encourage young designers and new motifs and use different textile bases. Cone Mills has a computer design base which has batik motifs that could be used in novel ways. Museums and universities play a key role. Do not forget the textile conservation issues. Bring artisans, marketers, consumers and other interested parties together to share ideas. Include hands-on and brainstorming sections. Develop a council to promote use of batik textiles not just to tourists and not just to locals but to both markets.

Slogans such as 'Batik ... the Soul of Java' would have to be revisited in order to reposition new products with a simple change of emphasis (i.e. 'Batik ... from the Souls of Java'). There is a subtle difference between the two but an important one. This could be used in a campaign complete with photographs of the souls who are designing and making *batik* including young and old souls.

References

Bennett, J. (1994), 'Hot Wax. An Exhibition of Australian Aboriginal Indonesian Batik', Museum and Art Gallery of the Northern Territory Travelling Exhibition.

Chevny, A.A. (1997), 'Sarung tenun Lamongan terancam punah', *Bisnis Indonesia*, 10 October, p. 19.

Connors, M. (1996), *Lao Textiles and Traditions*, Kuala Lumpur, Oxford University Press.

Crippen, K. (1994), 'Continuation and Change in "Geringsing" Double Ikat Weaving in Tenganan, Bali', presented at *Indonesian and Other Asian Textiles: A Common Heritage*, Jakarta, Indonesia.

Crippen, K. (1997), 'New Market Development: A Must for Asian Textile & Apparel Businesses', Singapore Conference, 1997.

Crippen, K. and Mulready, P. (1995), 'Textile Traditions and Quality of Life Concerns in Southeast Asia' in *Proceedings of the Fifth International Conference on Quality of Life and Marketing*, Williamsburg, Virginia, Academy of Marketing Science and QOL.

Crippen, K. and Vinning, G. (1996), Tapa Proposal – Asian Markets Research.

Elwood, A.P. and Lin, T.-b. (1983), *Tourism in Asia: The Economic Impact*, Singapore, National University of Singapore.

Herald, J. (1992), *World Crafts: A Celebration of Designs and Skills*, London, Letts.

Lucas, A. (1997), 'Alternative Trade: Aiming at fair share for Balinese crafts', *Jakarta Post*, 20 October, p. 7.

Lundberg, D.E. (1985), *The Tourist Business*, New York, Van Nostrand Reinhold Co.

Moss, L.A.G. (1994), *International Art Collecting, Tourism, and a Tribal Region in Indonesia*, Michael, P. (ed.), *Fragile Traditions: Indonesian Art In Jeopardy*, Taylor, Honolulu, University of Hawaii Press.

Samli, A.C. (1987), *Marketing and the Quality-of-Life Interface*, New York, Quorum Books.

Acknowledgements

Thanks to Textile Society of Hong Kong for supporting my trip to the conference.

Slides in this presentation are from the author's collection unless otherwise noted.

29 Modern Influences on East Sumbanese Textiles

KARIN SMEDJEBACKA

Sumba is generally considered to be an outstanding example of an area where the Indonesian textile tradition has reached its peak. The rich colours, the vivid motifs and the overall designs create a visually exciting art form. Also the skilfully employed techniques of both *ikat* and supplementary warp weave make the textiles exceptionally interesting.

The island of Sumba is one of the least developed parts of Indonesia. Fewer than half a million people live there, of which 165,000 inhabit the eastern part of the island. Also some Chinese and Arab families live on Sumba, and they control most of the businesses. The Chinese are the ones dominating the trade in textiles.

Textiles are produced also in the western part of Sumba, but this paper deals only with the textiles made in the eastern part, which are the ones that have attracted by far the most interest among outsiders.

More than 25 years ago, the anthropologist and art historian Marie Jeanne Adams studied the East Sumbanese textiles as well as different aspects of the society. In her writings she has repeatedly pointed out that the textiles reveal both the aesthetic and social order of the East Sumbanese society as well as the fundamental conceptions of value held by the people producing them. Adams (see, e.g., 1969 and 1971) also stresses that the textiles continue to play a vital role in the local culture.

In my own research I have examined how the textiles have changed from the time Adams studied them. I have also examined the influences of the commercial market and the social changes on the textiles. I spent the summer of 1995 doing field work in East Sumba, and will in this paper present some of the things I have learned.

The textile production in East Sumba is booming. This is surprising since a generally accepted assumption is that craft industries decline as traditional values disappear (e.g., Graburn cited in Geirnaert, 1989, p. 76; Taylor, 1994). On Sumba, traditional values are gradually disappearing, but so far this has

285

not led to a decrease of the overall production. By contrast, more textiles are being produced than ever before, albeit many of them of poor quality.

The increase in textile production started in the mid 1970s when Sumba, like so many other Indonesian islands, was drawn into the international art market (e.g., Taylor, 1994). Collecting art had become popular, and collectors were getting more and more interested in 'tribal art'. Sumba was one of the few remaining untapped areas which still had a wealth of objects that collectors wanted. Of special interest were the beautifully decorated East Sumbanese *ikat* textiles which also commanded high prices on the international art market. Consequently, an increasing number of art dealers started to arrive on Sumba (Moss, 1994, pp. 91–106).

At the same time the Indonesian government started to implement development plans on Sumba. As a result, the economic and social welfare of the East Sumbanese people began slowly to improve during the 1980s.

The demand for East Sumbanese textiles gradually expanded and this made it possible for the local weavers to raise their incomes by increasing their production. Since then, the production of textiles has become an important source of income, while at the same time the pressure to obtain money has been growing due to the process of development and modernization. Nowadays, people need money to pay for things like education, medical care and transportation.

The textiles are made in villages in five areas situated along the east and southeast coasts of the island. Each area has traditionally produced its own distinctive style, and some of them still do. Most of the textiles produced today are shipped off to holiday centres in Bali, Lombok, Sulawesi and Java or to factories in Bali and Java. These textiles are primarily sold to tourists or are destined for the international art market. Some of the textiles are, however, sold locally by pedlars or in the hotels and art shops, and a small amount is also produced for local use.

Unchanging Textiles

The East Sumbanese weavers produce two categories of textiles. One is made for the commercial market and the other one for local use. Both categories serve different but equally important functions in the community. Naturally, there is a difference in the attitude towards textiles produced purely for money compared to a piece made, for example, for a funeral. These attitudes influence the way in which a piece is produced and also the choice of materials used.

Generally speaking one can say that the attitudes towards the latter category of textiles have remained more or less unchanged. These textiles are used as clothing, and as gifts at weddings and funerals and are thus part of the traditional exchange system of the Sumbanese people.

Although the quantity of textiles produced for local use is small in comparison with the amount which is produced for commercial purposes, it is obvious that the locally woven textiles still constitute an integral part in the lives of the Sumbanese people. The quality of the textiles that are used in exchanges continues to be of great significance as indicators of value.

The most important type of East Sumbanese textiles being made today is the *hinggi*. *Hinggi* is the Sumbanese name for a big warp *ikat* blanket, which forms part of the men's traditional dress. Partly due to practical reasons the overall design of the blanket is composed of mirror image halves. Half of the motifs would otherwise be upside down when the *hinggi* is worn, as it is supposed to be either draped over the shoulder or wrapped around the hips. It is mainly these *hinggi* blankets which have attracted attention among visitors to the island. Today the great majority of textiles being woven are *hinggis* and they are also the type of textile which Sumba generally is associated with.

Traditional *hinggis* used locally require that natural dyes are used and that both the *ikat*ing and the weaving are done with much care. The quality of the these *hinggis* are consequently often high and therefore they tend to be expensive.

Nowadays, only a few weavers are skilled and devoted enough to make traditional *hinggis* and other textiles of high quality. These weavers still see their textiles as an important ingredient of the traditional way of life in East Sumba. If the current trend continues it will get harder and harder to find highly skilled weavers in Sumba in the future.

Changes in Work Patterns and Attitudes

In one of the weaving villages I met a family which provided me with a good example of how attitudes have changed. One weaver in her 70s spent all her time making high-quality designs, colours and weaving, which meant that it took her a long time to make each piece of textile. This did not matter to her, because she only wanted to produce textiles of high quality. Both of her daughters were, however, only interested in making as many textiles as possible in order to increase their incomes. They simply did not understand why their mother wanted to spend so much time on each piece of textile.

Not only attitudes, but also work patterns have changed. Before, the seasons used to determine when the various processes of making a piece of textile were carried out (e.g., Adams, 1969, pp. 79–80). This is no longer the case. Once the reasons for producing textiles started to change in the mid-1970s and quantity became the main objective, it was no longer feasible to work on a seasonal basis. While it used to take the weavers up to two and a half years to complete a good piece of *ikat* cloth, today, if more than six months are spent on making one, the local people think it has taken a very long time.

This change is closely linked with the building of roads in the 1980s. The new roads connected villages with the capital of East Sumba, which meant that the weavers in the outlying areas were able to buy materials such as threads and natural and synthetic dyes much more easily than before. With the roads, visitors started to arrive to the villages and some of them also bought textiles. Thus the weavers realized that by producing larger quantities of textiles they could earn more money. Consequently, no textile production is any longer carried out on a seasonal basis.

Work patterns have changed also in other ways. Previously, only women were allowed to take part in the production of textiles, but when the demand for East Sumbanese textiles started to grow in the mid-1970s, also men saw their chance to earn money in the expanding business. Since then the participation of men has increased significantly.

Before the impact of commercialization, it was not only the exclusive right of women to make textiles, but there also existed clear rules for which woman could do the weaving, *ikat*ing and dyeing. According to Adams (1971, p. 325), it was for example proper only for mature married women to do the weaving. Nowadays, you see many young girls and even men who weave. *Ikat*ing used to be the privilege of a few, and I was told that these women were held in especially high regard. Nowadays, anyone seems to practise *ikat*ing. Previously, dyeing was done only by a few initiated specialists, but with the introduction of synthetic dyes, dyeing can easily be carried out by anyone.

The female weavers have to a large degree also lost the artistic control they, according to Adams (1969, pp. 77, 128), previously had over the final product. Schneider (1987, p. 437) notes that this often happens when the production of a craft is commercialized. In East Sumba the final design of the majority of the textiles is nowadays determined by the commercial market, which in turn has led to that it is men and not women who create the desired designs. The women are left to take care of other less creative stages of the production.

Most of the taboos that once surrounded the production of textiles have been abolished. This has altered the powerful and prestigious position previously held by the female weavers. Hoskins (1989, pp. 143–4) points out that the taboos served to define boundaries between men and women and between those who had learned the mysteries of making a piece of cloth and those who had not. The social position of the female weavers therefore used to be clearly defined. As more and more people have taken up weaving, this position is getting increasingly ambiguous.

The textile researcher, Robyn Maxwell, has found that the attitudes of men towards textile production differ drastically from attitudes traditionally held by women. She points out that even where men are actively involved in the making of allegedly 'traditional' textiles, their products are intended principally for commercial markets. (Maxwell, 1990, pp. 404–5.) This is clearly the case with the textiles that the East Sumbanese men produce. Consequently, women have been influenced by the males' attitudes towards textiles, and the great majority of the female weavers therefore regard the production of textiles simply as a means to get money.

New Materials, Poorer Quality

The textiles we see today have to a large degree been shaped also by new materials. In the past, natural materials such as cotton, indigo and *gewang* palm leaves were used almost exclusively when making textiles. Now when bigger quantities are being produced, the majority of the textiles are synthetic in one way or the other. Synthetic threads are commonly used and so are synthetic dyes. I learned that even the *gewang* palm leaves used for tying in the threads or for *ikatting* were being replaced by strips of plastic.

The main reason for the use of synthetic materials is directly related to the demands of the commercial market. In order to increase the production volumes, the whole process of making a piece of textile has had to change. Using natural ingredients requires a lot of work and a lot of time. Consequently, the weavers have to use synthetic materials in order to speed up the production.

Another reason for the preference of synthetic materials is that the locally found natural resources no longer can support the growing production of textiles.

The main consequence of this trend is that the production of poor quality textiles has increased rapidly. The fact that most of the textiles currently produced for commercial purposes are of poor quality is not, however, surprising. According to several researchers (e.g., Schneider, 1987, p. 437), a prolonged contact with the outside world usually brings about a loss of traditional values which in turn is reflected in the lowering of the quality of the textiles, especially in the ones made for export.

On the other hand, commercialization has also made a small-scale production of high-quality textiles possible and, moreover, has revived the use of old traditional designs. This phenomenon has an obvious explanation. Textiles of high quality fetch high prices on the international art markets. Therefore, some of the most talented weavers have been able to spend all the time they need to make exquisite pieces. The phenomenon is important because it ensures that the knowledge and skills of how to achieve the highest form of the craft are being preserved.

Most of the textiles made for the commercial market are, nevertheless, of poor quality. Anyway, at least at present, the consumers seem to be willing to buy these textiles and this has made it possible also for less talented people to make a living by weaving. Furthermore, the growing textile business has attracted many of the younger members of society and they are now eagerly learning the craft. This is indeed a positive development.

New Designers and Designs

The commercial market has to a large degree influenced the choice of textiles the East Sumbanese weavers are producing. Many new innovations and alterations have been made during the last 20 years. They include the so-called modern *hinggi*, the bogus 'old' textiles and the use of non-Sumbanese design elements in combination with indigenous ones. These innovations have all been introduced by outsiders in order to make it easier to market the textiles.

Traditionally on Sumba there has not existed such a thing as a designer in the modern sense of the word. The patterns were handed down from one generation to another, from mother to daughter. The woman tying in the patterns was also the one who decided – or in other words designed – the final design. She was free to arrange fixed pattern elements according to her own taste (Adams, 1969, pp. 77, 128).

Now everything has changed. The turning point seems to have been in 1975, when the first so-called modern *hinggi* designs were made. The Chinese, who were controlling the textile trade, invented these new designs in order to create a broader market for the textiles and thus increase their own profits. At the same time men started to make designs, which previously was unthinkable.

This process led to the large-scale production of modern *hinggi*s, and nowadays they constitute one of the main types of textiles made for the commercial market.

A *hinggi* with a modern design is basically made in the same way as a traditional *hinggi*. They are, however, radically different because the modern ones are not composed of mirror image halves. This means that half of the motifs would be upside down if the *hinggi* was worn in the traditional way. The modern *hinggi*s can thus be used for example as hangings or bedcovers, but not as a traditional Sumbanese garment.

Towards the end of the 1970s, a great variety of modern designs started to appear. This trend has strengthened and today the overall designs are often spectacular, with many design elements put together in a dramatic way and with a strong display of colours. Most of the motifs are from Sumba, but some have also been taken from other islands such as Timor, Lombok and Sulawesi. The long, elongated, big human figures covering the whole length of the cloth are for example inspired by Hindu epic designs found on Lombok. Another popular motif is the Komodo dragon. Although big humans and dragons are traditional Sumbanese motifs, the ones which can be seen in these modern textiles have features not previously seen on the island. These new features can partly be explained by the borrowing of motifs from elsewhere and partly by the fact that some of the men who are now making the designs are outsiders, usually from neighbouring islands.

Another new development is that textiles are being made specifically to be sold as souvenirs. During a visit to Sumba, tourists usually take part in a funeral and also visit impressive ancestral houses and gravestones. Many of the modern *hinggi*s features all the main ingredients of the burial that the tourists have just seen. Another example is the *pasola* which is a mock cavalry battle fought annually in West Sumba. This is a major tourist attraction and the story about the *pasola* is therefore a popular theme in the textiles.

New designs can also be created from photographs. Sometimes even old designs are reintroduced in this way. I was for example told that in the late 1980s some tourists had arrived at one of the villages with a photograph of an old *hinggi* in which Queen Wilhelmina was the main design element. They wanted an exact copy to be made for them. Ever since, Queen Wilhelmina has been a popular motif in the textiles. Later on, more visitors have turned up with similar requests.

Textiles – a Part of Life

One of the general functions of textiles is to communicate identities and values (e.g., Schneider, 1987, pp. 412–6). On Sumba, the use of locally-produced textiles functions as a marker of ethnic identity by distinguishing the Sumbanese people from other Indonesians. Also the manner in which the Sumbanese men wear their *hinggi* is peculiarly Sumbanese and serves as an important sign of Sumbanese identity.

Monni Adams (1969, pp. 129–51) found that the design elements traditionally were related either to the weavers' own environment or were relevant to them in other ways, for example by expressing the history of the Sumbanese people or by depicting important rituals. Today, the weavers carry on this tradition when they are creating new design elements for textiles that they make for themselves. I came across this phenomenon in many of the villages. One woman was for example expressing her religious belief by rendering the motifs of a church and a Christian cross in the textile she was making.

The identity of the East Sumbanese is communicated also through the textiles made for the commercial market. Traditional design elements or variations of them are often used and these elements continue to carry symbolic meanings to the people. This became obvious when I discussed the textiles with people I met. The eagerness with which they explained to me what each design element depicted or stood for was striking.

The East Sumbanese are now proud of the fact that so many outsiders are buying their textiles, because this shows that the Sumbanese craft is appreciated. Equally important is the fact that it is the income from the textile production that enables them to take part in the ongoing process of modernization.

The Present and the Future

The outside world is just beginning to have an influence on the traditional way of life in East Sumba. This period of transition is therefore both complex and confusing.

The confusion is clearly reflected in the textiles. For example the designs of the modern *hinggi*s follow no clear rules and are often unbalanced and somehow disturbing to look at. The traditional *hinggi*s on the contrary have balanced designs and the weavers also follow strict rules when they structure the overall surface.

Many East Sumbanese are now eagerly trying to combine the ways of the modern world with the values of their own culture. They hold on to traditional values and beliefs, but at the same time they also want to be seen as being modern. Therefore things like education, Western clothing and status symbols such as televisions and motorcycles are becoming more and more important to them.

Laurence A.G. Moss (1994, pp. 116–9) has made suggestions on how to enable the weavers of the islands of the Lesser Sundas, of which Sumba is a part, to maintain their unique textile tradition. He thinks external assistance is needed and that it should be channelled through non-profit non-governmental organizations that have experience with cultural projects and, preferably, with the Lesser Sundas. He notes that most weavers need assistance in developing an accurate and timely understanding of the market, and that they also need a sympathetic marketing network. According to Moss (ibid., p. 118), this probably means eliminating most middlemen between the weavers and, initially, Bali and Jakarta.

These suggestions are in my opinion relevant to the present situation in East Sumba. The confusion that the current production of East Sumbanese textiles is displaying indicates that the weavers and the people that take care of the marketing should start to cooperate in an organized manner. I am sure assistance is needed in order to enable the weavers to compete in the harsh commercial markets of the future.

References

Adams, M.J. (1969), *System and meaning in East Sumba textile design: a study in traditional Indonesian art*, New Haven Connecticut, Cultural Report Series No. 16/Yale University/Southeast Asia Studies.

Adams, M.J. (1971), 'Work Patterns and Symbolic Structures in a Village Culture, East Sumba, Indonesia, Southeast Asia', *An International Quarterly*, 1(4), pp. 321–34.

Geirnaert, D. (1989), 'Textiles of West Sumba, Lively Renaissance of an Old Tradition' in M. Gittinger (ed.), *To Speak with Cloth, Studies in Indonesian Textiles*, Los Angeles, Regents of the University of California.

Hoskins, J. (1989), 'Why Do Ladies Sing the Blues? Indigo Dyeing, Cloth Production, and Gender Symbolism in Kodi' in A.B. Weiner and J. Schneider (eds), *Cloth and Human Experience*, Washington, Smithsonian Institution Press.

Maxwell, R. (1990), *Textiles of Southeast Asia: Tradition, Trade and Transformation*, Melbourne, Oxford University Press.

Moss, L.A.G. (1994), 'International Art Collecting, Tourism, and a Tribal Region in Indonesia' in P.M. Taylor (ed.), *Fragile Traditions: Indonesian Art in Jeopardy*, Honolulu, University of Hawaii Press.

Schneider, J. (1987), 'The Anthropology of Cloth', *Annual Revue of Anthropology*, 16, pp. 409–48.

Smedjebacka, K. (1996), 'Modern Influences on East Sumbanese Textiles – the impact of the commercial market on the product and its producer', unpublished MA thesis, Deptartment of Cultural Anthropology, University of Helsinki.

Taylor, P.M. (ed.) (1994), *Fragile Traditions: Indonesian Art in Jeopardy*, Honolulu, University of Hawaii Press.

30 Tourism and *Geringsing* Textiles in Bali: A Case Study from Tenganan

IR SULISTYAWATI, MS

Introduction

The purpose of this article is to describe the role of the unique *geringsing* cloth and the village of Tenganan, where it is produced, in attracting tourists and how this can be improved. Tenganan Pegeringsingan, besides being selected as a research subject for this article, has long been an interesting object of tourism, and now it is one of the most popularly visited villages in Bali. The total of tourist visits during the last three years, based on the numbers in the Tenganan village's guest book is as follows: at the end of 1994, the number of foreign tourists was 15,637 with domestic visitors of 2,152; in 1995, the number of foreign tourists was 28,422 with domestic visitors of 2,726; while in 1996, the number of foreign tourists was 33,380 with domestic visitors of 7,814. Based on the stated number of tourist visits to Tenganan, it can be concluded that there was a high increase, averaging 53.5 per cent, in the visits during the three year period. The interesting thing of those statistics is that there is an increasing interest in domestic tourists in visiting Tenganan Pegeringsingan village, representing an increase of 131.55 per cent.

Background

The combination of the uniqueness of its buildings, the village pattern, the villagers' way of life and customs are a specific interest for the foreign tourists who visit Tenganan. The Managing Director of Kuta Cemerlang Bali Raya (KCB) Tour & Travel stated that his staff consistently get orders for travel package to visit Tenganan for the foreign tourists, mainly from those who interested in culture. These tourists are mainly from Europe and the United

States of America, while the Asian foreign tourists normally have less interest in such issues (*Bali Post*, 25 January 1997, p. 8). The same opinion has been declared by the Managing Director of Bali Tour & Travel.

Weaving *geringsing* is the most popular handicraft in the village. Originally, this handicraft was used only as the traditional cloth of local villagers, but in fact it is traditionally also used by other villagers in Bali, because they consider that it has magical power. Because tourism developed early in the 1970s, *geringsing* cloth become known to foreigners; it started to be sold to outsiders, so that besides its art and magical values, it also has economic value. The same technique is found in Andhra and Pradesh, Orissa, India and in Japan. Exclusivity of *geringsing* cloth has become the trade mark of this village.

The thread is made of cotton and the dyeing process takes a very long time. In order to get a satisfactory result, these processes frequently take time, sometimes as long as two generations. The material of colour used to dye the cloth, is taken from plants around the village (Nusa Penida, Bugbug). The patterns of the cloth are well considered, using a perfect mathematical system, which is applied during the process of dyeing until the cloth is finished. Variation of the patterns shown in this *geringsing* cloth are really complex and unique (Seraya, 1981, p. 25).

The majesty of *geringsing* cloth in the past has been expressed by Urs Ramseyer in his book entitled *Balinese Textiles*, which records the use of *geringsing* cloth by Hayam Wuruk (King of Majapahit), as noted by Mpu Prapanca in *Nagarakertagama*. It is said that, the curtains of the carriage bearing Hayam Wuruk consisted of *geringsing lobheng lewih* and *laka*. The *Pararaton* chronicle tells that the first Majapahit king, Raden Wijaya, gave trousers of *geringsing* material to five of his soldiers, while the *Ranggalawe* says that the king gave sashes (*cawet geringsing*) to all of his warriors (Ramseyer, 1991, pp. 130–3).

Description of Tenganan Pegeringsingan Village

Tenganan Pegeringsingan is one of the traditional village communities which still exist in Bali. The location of Tenganan village is about 66 km east of Denpasar city. This village is surrounded by hills at the east, west and north sides so that they form a horseshoe-like fortification. It is thought that the word 'Tenganan' refers to its location in the middle of the hill (*tengahan*). In an *Ujung* inscription of the tenth century it is said that Tenganan village's name was originally Paneges, and it was located by the sea near Candi Dasa. Because of sea water erosion the villagers moved inward to the inland area (*ngatengahan*). Subsequently in its evolution, the word *ngatengahan* became Tenganan (Korn, 1960, pp. 307–10).

The word *pegeringsingan* has a direct connection with the *geringsing* cloth handicraft, which is the only handicraft in Bali where the process is done in this village. It is believed that the word *geringsing* is taken from two words *gering* and *sing* that are blended together. *Gering* means 'disease', and *sing* means 'no'. Thus, *geringsing* means 'immune from disease' and the Balinese community use it as a protection against sickness and black magic (Suwati, 1993, p. 41). *Pegeringsingan* means the place where the *geringsing* cloth is produced. Tenganan is definitely differentiated into two parts: Tenganan Pegeringsingan (*Banjar Kauh* and *Banjar Tengah*) with 300 inhabitants in 1997; and Tenganan Pande (*Banjar Pande*) with 214 inhabitants. The villagers venerate Dewa Indra as their all-mighty God. Tenganan Pande has a close relation to the existence of Tenganan Pegeringsingan because it is the place of exile for those who violate the village's traditional rules.

Gerinsing Cloth in Tradition

Formerly, people in Tenganan only used their own weaving material for their clothing, both for daily and ritual clothes, but now they have started to use other materials such as: sarong or *endek, songket,* or some are called *songket singepur* and batik cloth. In certain rituals such as *mabuang* dance, the other materials are combined with Tenganan's own production. The types of cloth produced by them are varied.

Types of *geringsing* are based on the width: *geringsing anteng,* which has the shortest width, for women using it as chest cloths and for men as a scarf in ceremonials; *geringsing patlikur* consists of *patlikur isi, patlikur cecempakan, patlikur lumbeng;* and *geringsing petangdasa* consists of *petang dasa lumbeng, petang dasa cecempakan, petang dasa wayang/kebo, petang dasa sanan empeg.* In general the function of *geringsing patlikur* and *petang dasa* for covering the chest (women) and *saput* (the outer part of traditional hip cloth).

Types of *geringsing* are also based on use: shawls for women, *geringsing wayang puteri, wayang kebo, gegonggangan;* for *anteng* (chest cloths), *geringsing cecempakan, patlikur isi, cemplong, tali dandan, teteledan, pepare;* for belts, *geringsing sanan empeg;* for *saput, geringsing wayang kebo, lubeng, cecempakan, cemplong, tali dandan;* and *sabuk tubuhan* for young men, *geringsing injekan siap, batun tuwung, setan pegat, dinding sigading, dinding ai* (Raka, 1978, p. 58). In general, *geringsing* should be used for *Bhatara* cloth in ritual ceremonies in temples. Besides *geringsing*, white and checked cloths are also used.

Geringsing cloth has many patterns, motifs and ornaments. Geometric pattern, motif, and ornament such as *tampak dara, swastika, tumpal* (mountain and *barong*'s teeth) are considered as having magical power and sacred value which can give prosperity and happiness. *Poleng* cloth is also considered as having power for protecting human beings from bad influences or danger. For these reasons, all of these textiles are used in ritual ceremonies (Melalatoa, 1977, pp. 71–2).

Out of the five traditional ceremonies (*Panca Yadnya*), three (*Dewa Yadnya, Pitra Yadnya, Manusa Yadnya*) very often use *geringsing* cloth. In *ngaben* ceremony, especially outside the Tenganan village, *wastra wali geringsing* (*wayang kebo, wayang putri*) are used to cover the top part of *wadah* or *bade*. The purpose is to protect the participants from accidents during the ceremony. In the second, *nelu bulanin – manusa yadnya* (the ceremony for a three-month baby), a child whose birth is affected by the death of his brothers needs this ceremony in order to ensure safety in his life. This ceremony is known as *wastra wali geringsing sanan empeg*, a symbol for protection from evil spirits. *Wastra wali geringsing cemplong* is also used for covering a pile of pillows during the tooth filing ceremony. Besides adding to the spectacle and the magnificence of the ceremony, this cloth also wards off evil. At the *mapedambel* ceremony, the last part of tooth filing ceremony, *geringsing wayang kebo mesemayut* is used (Seraya, 1981, pp. 18–31).

Geringsing as a Tourist Attraction

The uniqueness of *geringsing* is very often used as a trade mark of Tenganan village. The people of Tenganan, as well as travel agencies, use this characteristic for promoting Tenganan as one of the tourist attractions in Bali. It can be found in most brochures published by hotels or travel agencies, guide books, or even popular books which are published and sold to the public. On the Internet this kind of promotion is also available at http://warung1.com/balipg/geringsing.

Examination of publications from 93 travel agencies in Bali (Daftar Biro Perjalanan Wisata, 1996) has revealed the ones that promote *geringsing* cloth. The following quotations were taken from these diverse sources:

Illustration 30.1a   *Geringsing* cloth hanging inside houses in the
village of Tenganan, to attract tourists' attention.
Geometric and abstract floral motifs in most of the
clothes in Illustration (a) are repeated horizontally,
vertically and diagonally over the cloth. These
geometric patterns and motifs are considered to
have magical power and sacred value

Illustration 30.1b *Geringsing* cloth hanging inside houses in the village of Tenganan, to attract tourists' attention

Illustration 30.2   Kepala Desa Adat Tengganan wearing the *geringsing* cloth

Tenganan features wonderful fabrics, one of which is the renowned geringsing double ikat weave ikat cloth (Nusa Dua Bali Tours & Travel).

This Village is famous for its 'double ikat' woven material called 'gringsing', which is supposed to protect the wearer with magic power (Brochure Astina Tours & Travel).

The inhabitants of Tenganan spend most of their time preparing offerings and producing the textiles for which it is famous. The peculiar double ikat commonly known in Bali as geringsing is woven in only two other places in the world; Patola in India, its likely place of origin, and Guatemala ('Cloth Culture', *Garuda Monthly Magazine*, July 1997).

Kain geringsing dikagumi para kolektor kain yang serius sebagai maha karya kriya tenun, kain keramat ini menakjubkan ('Kain Para Dewa', *Official Guide Book of Sempati Air,* Cakrawala-Bali, 1995).

The women of this village weave the famous 'flaming' cloth, kamben geringsing, which supposedly has the power to immunize the wearer against evil vibration. The gringsing cloth tends to shimmer in the sunlight and it is worth a visit just to see the people dressed in this amazing fabric (*Insight Guides BALI*, 1995).

A magical cloth known as kamben gringsing is woven in Tenganan – a person wearing it is said to be protected against black magic (*Travel Survival Kit INDONESIA*, 1986).

Young women, dressed in headdresses of gold flowers and sarongs of the village's double ikat, or geringsing, dance in tight formation. ... Only a few women still make geringsing, which is endowed with mystical powers and created in what must be the most difficult dyeing procedure yet devised (Charle, 1990, pp. 136–8).

Certainly one of the rarest weaving techniques in the world is practised in Tenganan, a traditional native Balinese village in eastern ... Nowhere else in Indonesia is this 'double ikat' practised and only about 15 women know how to weave it (Dalton, 1990, p. 85).

The method for making geringsing cloth is so time-consuming that very little is made any more, and, if any can be found to be on sale it is almost prohibitively expensive (Eiseman Jr, 1990, p. 237).

It is clear from these above quotations that tourism actors and cultural commentators in Bali, directly or indirectly, have promoted *geringsing* cloth intensively around the world. But for the Tenganan people, making *geringsing* remains an important part of their culture.

Interestingly, official tourism bodies in Indonesia often overlook this craft: *Informasi Obyek dan Daya Tarik Wisata di Bali*, published by Dinas Pariwisata Pemda Tingkat I Bali in cooperation with 'PUTRI' Tingkat I Bali, promotes neither *geringsing cloth* nor Tenganan Village as tourist attractions.

The Result of Tourists' and Weavers' Questionnaires and Interviews with Tourist Guides

A small survey of 20 tourists in Tenganan revealed that the following countries origin: France (5), Italy (1), Holland (4), Denmark (4), Spain (2), Australia (1), USA (2), and Canada (1). It can be concluded that most tourists who come to Tenganan are from Europe. Generally tourists who come to Tenganan are individualists or Family Individual Travel (FIT).

Twelve tourists (60 per cent) came to Tenganan because of the uniqueness of the village; six tourists (30 per cent) because of the uniqueness of Tenganan people's custom; six tourists (30 per cent) because they wanted to satisfy their curiosity; the others because of other reasons such as buying a basket, visiting friends, etc. So, of this small sample *geringsing* cloth is not the prime attraction for tourists to come to Tenganan.

When they were asked whether they knew *geringsing* before they came to Tenganan, 19 respondents said 'no' and only one respondent said 'yes'. This condition shows that promotion of *geringsing* through brochures, guide books, or other books is not yet capable of creating an attractive image of cloth culture.

The questionnaires were distributed when the tourists had finished visiting Tenganan village. Of 20 respondents, 11 (55 per cent) declared that they were interested and nine (45 per cent) were not interested in *geringsing*. The reasons of those who are interested were: five respondents (45.4 per cent) because of the uniqueness and the beauty of geringsing motif; four (36.4 per cent) because of the uniqueness of its technique; two (18.2 per cent) because of the beauty and the magical power of its colour; one respondent did not mention any reason. The tourists became interested in the *geringsing* cloth after they had seen it for themselves.

When they were asked whether or not they possessed *geringsing* cloth, only one respondent said 'yes'. Those who did not have this kind of cloth gave the following reasons: nine (47.4 per cent) found it too expensive; two

(10.5 per cent) could not find suitable motifs; one (5.3 per cent) thought it not useful; and six (31.6 per cent) gave different reasons such as '*geringsing* cloth is only for the local people', 'we do not know much about *geringsing*', and 'it is not suitable as a hobby'.

A small survey of designers, weavers, and dyers has also been undertaken. It revealed that of 20 respondents: none of them was younger than or 15 years; 16–25 years = 10 (50 per cent); 26–35 years = 3 (15 per cent); 36–45 years = 3 (15 per cent); and older than 46 years = 4 (20 per cent). From the data it can be concluded that there has been a process of skill regeneration in *geringsing* weaving; the number of young weavers is over than 50 per cent. One of young weavers (26–35 years old) has even mastered all procedures in weaving a *geringsing* cloth.

The survey revealed the professions of the respondents as follows: eight (40 per cent) are *ikat* binders; five (25 per cent) are dyers; 15 (75 per cent) are weavers; and six (30 per cent) are designers. When the statistics are added together, the total is 170 per cent, showing that many weavers have more than one profession.

The respondents' income per month is as follows: none is lower than Rp 150,000; Rp 150,000 – Rp 300,000 = 3 (15 per cent); Rp 300,000 – Rp 450,000 = 8 (40 per cent); Rp 450,000 – Rp 600,000 = 4 (20 per cent); higher than Rp 600,000 = 5 (25 per cent). Based on the tabulation, the respondents with incomes higher than Rp 600,000 are those whose professions are as designers and at the same time as producers. Only one of the designers' income was lower than Rp 600,000; she was less productive because of her advanced age. The survey also shows that even those who have one profession are able to earn Rp 450,000 – Rp 600,000, depending on their ability in dealing with the difficult process of making *geringsing* cloth, particularly in weaving *geringsing wayang*.

Based on the weavers' evidence about the number of tourists who visit their places of work in a month none of the tourists visited them longer than 45 minutes. Based on the tabulation it is revealed that all tourists who visit the weavers' places of work ask the weavers about *geringsing*.

The comments given by tourist guides during the interviews provide some tentative insights. Two English-speaking guides said that American tourists have no interest in *geringsing* as they are more interested in Tenganan's ceremonies. Four French-speaking guides commented that some French tourists are interested in it. Two German-speaking guides admitted that the Germans are not only interested in the 'double *ikat*' but quite a lot of them bought it. Two Spanish-speaking guides and a Japanese-speaking guide said that the tourists visit Tenganan usually because it is included in the itinerary of the tour package. A guide handling Korean and Taiwanese said that his clients have no interest in visiting the village. Thus, it can be concluded that the role of textiles, in this case *tenun ikat*, is relatively small in supporting tourism in Bali.

Conclusion and Suggestion

These results are inconclusive, but suggest that the role of *geringsing* cloth is not significant in attracting tourists to Tenganan village. The information about *geringsing* in brochures and books also do not much influence the visits of tourists to Tenganan. *Geringsing* cloth represents a small portion of the wider culture of Bali. Most of the tourists said that they come to Tenganan village because they are interested in the uniqueness of the old village and its customs.

For Tenganan villagers, the role of *geringsing* is mainly job creation, especially for the local women. It also helps to conserve local customs and cloth culture and at the same time provides more money for supporting local families.

It is suggested that the Tenganan people should be more active in promoting their traditional cloth and customs through national and international exhibitions, as was the case in July–August 1997 in Tokyo, Japan. Reliance on books and brochures written and published by foreigners is too limiting.

References

Bailitis, E. (1997), *Cloth Culture. Garuda: The Official In Flight Magazine of Garuda Indonesia*,Vol. 5, Brisbane, Garuda Indonesia.

Charle, S. (1990), *Bali: Collins Illustrated Guide*, Great Britain, The Guidebook Company Ltd.

Cumings, J. et al. (1986), *Travel Survival Kit INDONESIA*, 2nd end, Hawthorn, Lonely Planet Publications.

Dalton, B. (1990), *Bali Handbook*, USA, Moon Publication.

Deparpostel (1996), *Daftar Biro Perjalanan Wisata Tahun 1996*, Denpasar, Deparpostel-Kanwil Bali.

Eiseman Jr, F.B. (1990), *Bali Sekala and Niskala*, Vol II, Singapore, Periplus Edition (HK).

Ensiklopedi Nasional Indonesia 16 Ta-Tz (1991), *Tenun Ikat*, Jakarta, PT, Cipta Adi Pustaka.

Gillow, J. and Barnard, N. (1996), *Traditional Indian Textiles*, London, Thames and Hudson.

Hauser-Schaublin, B., Nabbholz-Kartaschoff, M.-L. and Ramseyer, U. (1977), *Balinese Textiles*, Singapore, Periplus Editions.

Kartiwa, S. (1986), *Kain Songket Indonesia (Songket Weaving In Indonesia)*, Jakarta, Djambatan.

Kartiwa, S. (1987), *Tenun Ikat Indonesia (Indonesian Ikats)*, Jakarta, Djambatan.

Korn, V.E. (1960), *Bali, Studies in Life, Thought and Ritual*, The Hague, van Hoeve.

Melalatoa (ed.) (1977), *Adat Istiadat Daerah Bali*, Jakarta, Proyek Pengembangan Media Kebudayaan Depdikbud.

Oey, E. (ed.) (1991), *Bali Island of The Gods*, Singapore, Periplus Editions.

Raka Dherana, T. (1976), *Sekilas tentang Tenganan Pegeringsingan*, Denpasar, Bagian Penerbitan Fak. Hukum & Pengetahuan Masyarakat Unud.

Ramseyer, U. (1984), *Cloting Ritual and Society in Tenganan Pegeringsingan (Bali)*, Basel, den Verhandlungen der Naturforschenden Gesellschaft.

Reyes, Elizabeth V. et al. (1987), *The Travel Library, Bali*, Singapore, Tein Wah Press.

Seraya, I .M. (1981), *Wastra Wali*, Denpasar, Proyek Permuseuman Museum Bali.

Team Feasibility Study (1979), *Naskah Feasibility Study Desa Tenganan*, Jakarta, Proyek Sasana Budaya.

Wiratanuningrat, P. (ed.) (1995), *Insight Guides BALI Baru*, Singapore, Hasan M. Soedjono.

Interviews with informants, July–September 1997:

The Head of Tenganan Village	:	Dr Nengah Wartawan (35 years).
Priest	:	Mangku Widya (57 years).
Guides	:	Drs Matt. Oka Wirawan, Ketut Jaman, S.S.
Designers	:	Nengah Nuri (67 years), Nengah Rantin (75 years), Ni Wayan Rantis (61 years).
Weavers	:	Luh Sutini (25 years), Nengah Rawit (38 years), Nengah Surtini (21 years), Ni Landri (47 years), Wayan Rusni (32 years), Nengah Suastini (33 years).
Dyers	:	Ketut Rentig, Nengah Rusni.
Tiers	:	Luh Yudiani (25 years).
Others	:	Ketut Tantra (44 years), Nengah Timur (37 years).

31 *Endek* and its Role in Tourism Development in Bali

TRI BUDHI SASTRIO

All women's dresses, in every age and country, are merely variations on the eternal struggle between the admitted desire to dress and the unadmitted desire to undress (Lin Yutang, quoted in *Ladies' Home Journal*).

Introduction

The word *textile* (from the Latin *texere*, 'to weave') originally meant a fabric made from woven fibres. Today the term signifies any of a vast number of fabrics produced by weaving, knitting, felting, and other techniques such as *ikat technique*. The term *ikat* itself literally means 'to bind'. This is a technique of ornamenting a fabric. Thin fibres are wound around a thread before weaving, to protect these bound parts from absorbing the dye into which the threads are dipped.

Three kinds of *ikat* may be distinguished: *ikat of the warp, ikat of the weft,* and *ikat of all the threads*. Those three kinds of *ikat* can be produced either by using *floating weft technique* or *pilih* (*pilih* = 'to select') *technique*. In the first technique the thread forming the pattern is woven in simultaneously with the thread which holds the fabric together (as a cross-thread). The latter is the ordinary weft thread. For this purpose the warp threads are provided with shed sticks. Meanwhile, the second technique actually closely resembles the first technique except that in this case the thread forming the pattern is passed through the threads of the warp where necessary with the aid of a long spool (Wagner, 1959, pp. 246–7).

Significance of Textiles

As in the custom the world over, textiles are used to make articles of clothing, in particular for clothes worn on festive occasions both sad and gay, and for ceremonies, as well as clothes worn to identify the wearer as a member of a certain class or as holder of a certain rank. But what makes these *tenun ikat* so immensely significant is the special part they play in many ceremonies and customs in the life of the Indonesian peoples. Certain fabrics have a sacral and/or ceremonial function to fulfil at birth and death, at important events such as circumcision and filing the teeth, in custom connected with marriage, as well as in certain rites observed when planting rice.

The Indonesian envisages not only the world of nature as male and female, but also the whole cosmos, including material objects. Sun and heaven are the masculine counterparts of moon and earth. Textiles represent the *female* element, whereas weapons (spears, kris), on the other hand, represent the male element; when both are combined they symbolize completeness. The carrying of flags and pennons, which is generally considered as merely a festive ornament, or a means of expressing joy on some special occasion, thus acquires a symbolic significance in the truest sense of the word.

Such significance attaches to the textiles not only of those peoples relatively unaffected by later cultural influences, but also to those strongly permeated by Hinduism (e.g. on Java and Bali). On these islands certain textiles are still especially esteemed, in particular the old calico cloths produced by the tie-dyeing process, which are deemed to possess special magic powers (ibid., pp. 51–2).

Traditional Cloths: Batik and *Endek*

Batik is a wax-resist dyeing process for fabrics used by the Javanese for about 1,000 years. The technique involves applying molten wax to those areas of the cloth that remain undyed. When the waxed fabric is submerged in a dye bath, the unwaxed areas are coloured, and the wax-coated areas repel or resist colouring.

Traditional batiks are made in the following manner. The fabric, usually cotton, is washed and dried several times then a diluted starch solution is washed in as a sizing. Next, the fabric is beaten to smooth it out and to prepare the fibres for dyeing, meanwhile the design is drawn on the fabric. Following the design, the wax pattern is drawn on the fabric with a spouted applicator called a *canting* or printed on with a copper stamp called a *cap*. Next, the fabric is dyed and, when dry, the fabric is either scraped or boiled to remove the wax. The waxing and dyeing process is repeated for each colour in the finished batik.

One does not have to be an expert to know there are numerous types of batik. There are differences in material, such as *cotton* and *silk*, and recently synthetic materials, as well as *linen* and *wool*. There are differences in technique, such as the *hand-drawn, printed*, or a combination of the two. There is the printed material, similar to ordinary material only using traditional batik designs.

However, it takes an expert's eye to differentiate between the batik originating from different regions and even more discerning knowledge to distinguish between works of different batik artists. The same thing is also happened to the *traditional cloths* in Bali (i.e. *songket, geringsing, endek*, and *batik Bali*).

Endek is a *Balinese weft ikat cloth*. It is a woven, tie-dyed weft cloth. Balinese *endek* is tied and dyed before the threads are woven into cloth. In preparing *endek*, the weft or cross threads are dyed; the warp, the threads that are initially strung on the loom, is left in a solid colour. To prepare the pattern, the weft threads are temporarily strung on a frame and workers use strips of plastic tape to 'tie' a pattern into the threads. The threads are taped off in bunches, and then the threads are removed from the frame and soaked in vats of dye. They are dried, the tape is removed, and then the thread is spun onto the shuttle. When the dyed threads are woven into a loom set up with a solid warp, the design reappears. Because the warp threads are taped off in bunches, and because perfect registration of the design is impossible, the finished *endek* pattern has an attractive, fuzzy-edged look (Eiseman, 1990, p. 232).

In the past, *endek* cloths were generally the prerogative of the princely courts and aristocratic families with regard to both production and use. These cloths have since undergone a process of *democratization*, so that personal distinctions can no longer be made on the basis of textiles alone. They are also social badges (Hauser-Schaublin, 1997, p. 12).

In 1990s Bali, *endek* is created and produced not only for the tourists who come to Bali but also for overseas markets. Very often, the design is created by collaboration between foreigners and Balinese, while still maintaining the traditional patterns. But the culture, the way of life, and the religious activities of Balinese are too strong to compromise even the pattern of their *endek*, so the buyers are required to adjust their taste.

Endek for the Balinese

Endek for the Balinese, like *batik* for the Javanese, is considered as a part of their distinctive culture. It evokes a rich tapestry of historical and ritual association, as well as a tourist attraction, an important element in pushing tourism development. However, at least before this century in Bali, the textile-dyers, like butchers and potters, were still considered to be impure. They command respondingly low respect. It is only the low *sudra* who maintain their livelihood by carrying out these income-generating activities.

Bali has been changing since Indonesian independence, and the rules of the old Hindu upper orders, the *caturwangsa (brahmana, satria, wesia,* and *sudra),* are no longer the only guide to social interaction. Modern life tends to diminish those social strata, and then change them into something new, the strata based on professionalism. The textile-dyers are no longer considered as 'low-professionals'. These professions are one pillar for pushing tourism development. These professions are important not only for maintaining the cultural identity of Balinese but also for maintaining the cultural identity of Indonesia. Some *endek* producers in Sidemen strongly support this statement.

Clothing Industry in Bali

The clothing industry contributed 61.3 per cent to the total export value of the province Bali in 1990 (*Statistik Bali: Statistical Yearbook of Bali,* 1991, p. 360). This percentage has increased significantly during the last five years. Profiting from direct tourist contact, it superseded coffee in 1981 as the most profitable export product. In Bali in the 1990s, fashion is created and produced not only for Indonesia's large cities but also for overseas markets.

The designs are created by collaboration between foreigners and Indonesians and often have little to do with traditional patterns. Production, however, exhibits a typically Balinese component: it could only have originated in the touristically accessible south Bali. The international airport guarantees connections with foreign countries and jobs are welcome, though often poorly paid (Hobart, 1996, pp. 219–20).

Tourism and its Impact

The growth of tourism has had far-reaching social and economic effects on the island of Bali. It has provided new employment opportunities, not only in hotels but also in arts and crafts, in entertainment, and in travel agencies. This prospect of employment is one of the many factors contributing to population growth in the major tourist areas. However, at the same time, the tourist boom has resulted in rising land prices, land speculation, and the conversion of land from agricultural to non-agricultural uses. More importantly, mass tourism has introduced organizational changes in Balinese tourism. There has been a transition of ownership out of Balinese hands, and the Balinese response to tourism is being increasingly orchestrated by outsiders, mainly Indonesians from Jakarta and the transnational corporations.

The impact of tourism at the district level has been uneven, as the regional development plans admit. At the *banjar* level, both McKean (1979) and Udayana University (1979) found no uniform procedure for the distribution of income derived mainly from cultural performance for tourists.

For example, during the 1930s, the tradition of *endek* production and use began to detach itself from the closed world of the courts and underwent a renewal in many villages in Bali. Weavers began to make simple *endek* materials from hand-spun local cottons or from factory-produced and patterned yarns on traditional *cag-cag* looms.

After independence, this development proceeded at an explosive rate. During the 1950s, the first large workshop were set up in Gianyar and these have now grown into important factories. In the 1970s, workshops large and small mushroomed all over Bali, in Sidemen, in the Singaraja area, in Sampalan near Klungkung, and in the neighbourhood of Negara (Jembrana, West Bali). By 1989–90 there were 160 commercial producers in Bali employing a total of 10,042 people. The production of checked, striped, plain, and *endek* materials from cotton as well as man-made fibres and silk had by this time burgeoned to an average of 188,000 metres per month (Hauser-Schaublin, 1997, pp. 19–20). This impressive number (188,000 metres per month) was declining in the 1990s, but the quality of the products was increasing significantly. What are the reasons behind this phenomenon?

The process of modernization has been marked by such decisive innovations as the application of new and more efficient winding and warping methods, the use of more convenient tying materials, the introduction of fast-acting synthetic dyestuffs, and a change-over from the traditional *cag-cag* or backstrap loom to the new ATBM loom.

Sidemen as a Case Study

In brief, most *endek* producers in Sidemen claimed that before the 1970s not many visitors came to their village. Not only because their village was considered to be a remote village, but also because there is nothing unique for the visitors. But after the 1970s, when *endek*'s production started in this tiny village, this condition changed completely. Visitors and foreigners alike came in droves and bought *endek*.

The result is the entrepreneurs of *endek* and their employees have prospered. Using the words of Brigitta Hauser-Schaublin (1997) in describing an *endek*'s producer in Sidemen we got a more vivid picture about one of them.

> With no land of his own to begin with, he has been able to buy rice and vegetable fields over the course of the years, and to employ others to farm them for him. He owns two rice mills in the village, two house compounds, and a house in a suburb of Denpasar where he goes in his car on business or to visit his older children, who are receiving higher education there. However, as we have noted, the manufacturers shave their price very finely and competition between them is fierce. A single mistake in color, pattern or material, and all that he possesses could be at stake.

The competition between them is in fact really fierce but the number of visitors come to their village to see and to buy *endek* is still high. Balinese buy the *endek* for *sarungs*, a waist wrap worn by men, and for *kamben*, a cloth worn by women and, on ceremonial occasions, by men. Visitors prefer to buy *endek* that is made into shirts, blouses, and dresses.

The chart below shows some types of cloths made of *endek* worn by Balinese men and women.

	Men	**Women**
Cloths for head	*destar*	*tengkuluk/gelungan*
Cloths for body	*kampuh*	*senteng/stagen-lamak*
Cloths for lower part	*wastra*	*wastra istri/sinjang/tapih*

Endek and its uniqueness are part of so many factors which influence positively tourism development in Bali. But what about its role in the future? So far nobody knows, but of one thing we are sure: as long as the Balinese still live in their culture and put the practice of their culture into the way of their life, the role of *endek* will be still great in the future as already shown in the past.

References

Bandem, I.M. (1996), *Wastra Bali: Makna Simbolis Kain Bali*, Denpasar, Hartanto Art Books.

Covarrubias, M. (1937), *Island of Bali*, New York, A. Knopf.

Eiseman, F.B. (1990), *Bali: Sekala and Niskala, Vol. 1, Essays on Religion, Vol. 2, Essays on Society, Tradition and Craft, Ritual and Art*, Hong Kong, Periplus.

Hamzuri (1981), *Classical Batik (Batik Klasik)*, Jakarta, Djambatan.

Hauser-Schaublin, B., Nabbholz-Kartaschoff, M.-L. and Ramseyer, U. (1991), *Balinese Textiles*, London, British Museum Press.

Hobart, A., Ramseyer, U. and Leemann, A. (1996), *The Peoples of Bali*, Oxford, Blackwell.

Puja, G. dan T.R.S. (1997), *Manawa Dharmacastra*, Jakarta, Hanuman Sakti.

Wagner, F.A. (1959), *The Art of An Island Group*, New York, Crown Publisher, Inc.

Wertheim, W.F. (1960 [1907]), *Bali: Studies in Life, Thought, and Ritual*, with an introduction by J.L. Swellengrebel, The Hague, Van Hoeve.

32 Textile Production, Tourism and the Internet – The Austrian Experience

AMIN TJOA AND ROLAND R. WAGNER

Shall we rouse the night-owl in a catch,
that will draw three souls out of one weaver (William Shakespeare)

Introduction

In the last decade much effort has been invested in exploring the role of textiles as an important ingredient in special-interest tourism in Austria. The aim of this chapter is to give an overview of the different tourism marketing strategies which use textile art to achieve their goals. The spectrum of initiators' actions is very broad in that it covers local and regional tourist organizations, local museums, federal museums, etc. The Austrian Tourist Information System endeavours to present a broad range of objectives for cultural and heritage tourism on the Web. Art in general and also textile art feature prominently in the information presented to the clientele reachable via the Web.

The efforts of the rural areas of Mühlviertel will be described in detail. This region between the Bohemian Forest and the Danube (e.g. the village of Haslach) is famous for its activities concerning the promotion of the heritage of linen production. Due to the growing demand for linen between the sixteenth and eighteenth centuries, all continental Europe bought the woven products of this region. The triumphant advance of cotton and the industrial revolution put an end to this development. Nevertheless, the weaving tradition continues in the villages of Mühlviertel. In this area we can find both the history of weaving and the newest developments in the museums, art galleries, and ateliers. Every summer the village of Haslach organizes the *Textile Kultur Haslach*, an event which is attended by many tourists interested in the creativity and variety of textile handicraft in Mühlviertel. Also, different workshops on weaving are organized with great success by the tourist board. The Web presentation of this region is an important instrument in direct electronic marketing.

Another example of these initiatives on weaving can be seen in the region of Steiermark. Spinning, knitting and weaving of wool play an important role in the promotion of this region by the local tourist board. Lamb's wool of this region is coloured with traditional colours made of plants, barks and roots. The production of handicraft souvenirs is an important economic activity in this region. Some villages in this area (e.g. Kapellen) combine the idea of museums and the selling of home and hand made quality products for tourists in an ideal manner.

The Museum of Ethnology, Vienna is also increasingly concentrating on exhibitions of textile arts. The recently organized exhibition of Indonesian textile art, which provided an in-depth view of the textile products and creations of the different regions of Indonesia, was a major cultural attraction for the Austrians. In addition, the event also attracted many tourists visiting Austria. This event was also publicized on the Web.

Linen Production in Mühlviertel

As early as the thirteenth century in Upper Austria, and especially in Mühlviertel, wide areas were cultivated with flax (*Linum usitatissimum*), a plant of the family *Linaceae*. The flax plant is primarily cultivated for its fibre, from which linen yarn and fabric are made. Flax is one of the oldest textile fibres. Evidence of its use has been found in the prehistoric lake dwellings of Switzerland. Fine linen fabrics, indicating a high degree of skill, have been discovered in ancient Egyptian tombs. The oldest Austrian archeological discovery of linen-seeds has been found in *Lauriacum* (Upper Austria), dating from the Roman age.

Linen is valued for its strength, lustre, durability, and moisture absorbency. It is resistant to attack from micro-organisms, and its smooth surface repels soil. It is stronger than cotton, dries more quickly, and is more slowly affected by exposure of sunlight. One disadvantage of linen is its low elasticity and its hard, smooth texture which tends to wrinkle. Fine grades of linen are made into woven fabrics and laces (see *Britannica Online*).

The history of linen production in Mühlviertel is associated with prosperity until the middle of the nineteenth century. The second half of the nineteenth century is overshadowed by competition from cotton and Irish linen, which undercut local prices from the 1880s onwards. The low price of the exports amounted to dumping in Austrian eyes and ruined many small weaving enterprises. During World War I and shortly after, the agricultural area of flax cultivation decreased from 2,930 ha to 785 ha. In 1927, due to governmental protection measures for cotton and imports, foreign competitors were largely eliminated from the Austrian market. New domestic factories were founded.

Hand weaving still survives within the farms of the Mühlviertel region. One of the most famous artists in the field of hand weaving in this time was Wenzel Geretschläger (1852–1945) who was well known throughout Europe for his woven works of art. His masterpiece 'Hunter's Wedding' can be found in several museums and has been exported to many countries. His son Josef Geretschläger has produced more than 2,800 pieces of 'Hunter's Wedding' on a Jacquard loom.

In the 1950s and 1960s the production of linen drastically declined. Important flax processing manufacturers (e.g. Fa. Haberkorn & Co.) had to stop production. The last large order for linen production was an order for uniforms for the Austrian army in 1959, which aimed to stimulate domestic linen production. In the 1960s there was little cultivation of flax in Mühlviertel due to the use of synthetic fibres by the manufacturers.

The Recovery of Linen Production and Tourism

In the 1970s the demand for natural fibres increased due to the increasing interest in cultural heritage and the growth of environmentalism. In 1977 the Austrian Federal Ministry of Science and Research and the Austrian Chamber of Commerce launched a research project to provide an inventory of weaving articles produced in the Mühlviertel region. Despite the efforts invested in this project it was not possible to reach an understanding of the entrepreneurs of this area regarding a unique product and marketing concept. One important aspect in this time was also the huge risk of converting to natural fibres (i.e. flax). In the 1980s flax was again cultivated on a larger scale in Mühlviertel, after an absence of nearly 20 years. This began with some controlled experiments which were very encouraging. In the 1990s flax is harvested with modern machines for further processing in Knittelfeld and Rastenfeld.

For the colouring process, traditional colours, especially made of plants (e.g. *indigofera tinctoria*), are used for the blue printed patterns. In the nineteenth century 17 dye-houses largely engaged in this kind of printing. Some of them were producing in this manner until the late 1950s. *Zötl*, for example, was producing from the eighteenth century until 1956. Since the

1970s, natural traditional colours have been increasingly used in Mühlviertel. The renaissance of the weaving industry in Mühlviertel is characterized by traditional production and modernized automated production. For modern production in the medium enterprises high-tech methods such as computer integrated manufacturing plays a dominant role. At present 85 per cent of the fibres processed in these companies are natural fibres, especially flax. The revival of the linen industry is also accompanied by a shift from household furnishing to apparel. In the 1960s only 1 per cent of the linen production was used for apparel, whereas now it reaches nearly 50 per cent. Currently 500 weavers are employed in 18 small and medium-sized enterprises. In this region the linen and weaving industry is regaining its traditional role despite the fact that rationalization significantly reduces the total number of people employed.

With the help of the Austrian Chamber of Commerce a continuous observation of the fashion market is initiated, making it possible for the small and medium-sized enterprises to react swiftly to trends in the fashion area. The use of innovative natural materials and the creation of new articles are possible because of the establishment of specialized dye production and design departments, which are responsible for the flexible adaptation to the demand of the market.

Another important factor for the revival is the modernization of the vocational school of weaving which was founded in 1883 and which has – with an interruption of 20 years at the beginning of this century – continuously educated specialists in this area. Now this school guarantees the existence of a next generation of skilled workers and specialists who are trained in traditional skills, as well as the modern techniques of textile design and production (e.g. CAD, CIM, etc.).

In 1990 the weavers of Mühlviertel started an initiative for the establishment of a tourist route through the different tourist attractions which are linked with the weaving industry in this region (*Mühlviertler Weberstrasse* – Mühlviertel weaver's route). The project team consists of representatives of the Chamber of Commerce, a representative of the vocational school, tourism experts, experts on culture and the arts. The project leader of these 10 experts was one of the directors of the weaving enterprises.

In 1992 the touristic *Mühlviertler Weberstrasse* was officially established and financed with seedcorn money from the weaving industry of the region. The *Mühlviertler Weberstrasse* consists of 16 towns and villages in Austria and one town across the border in Germany. In all these places the weaving industry and related attractions (e.g. museums) play an important role. Groups of tourists or individuals are able to visit these attractions systematically.

This tourist marketing initiative is accompanied by the following measures (see Heindl, 1992):

a) research on the history of textiles in this region;

b) demonstration of the production chain from the cultivation of flax to the sale of the textile products;

c) stimuli for the innovation of future textile products;

d) strengthening of the consciousness of regional roots;

e) research on how the *Mühlviertler Weberstrasse* could contribute to the economic revival of the region;

f) the contribution of the *Mühlviertler Weberstrasse* to the cultural and economic development of the region.

After five years of experience one can conclude that the recent development of tourism in this region is tightly or even totally linked to the development of the *Mühlviertler Weberstrasse*. In other words, there would be hardly any significant tourism in this region if the *Mühlviertler Weberstrasse* did not exist. Every touristic activity or initiative is connected in some way to textile production (e.g. the weaving museum in Haslach, souvenir shops in factories, etc.). Of course the *Mühlviertler Weberstrasse* initiative has also integrated other attractions of the region into its programme (e.g. the monastery in Schlägl and the scenic Moldau River).

The following examples illustrate the different activities which contribute to the success story of tourism in this region:

a) the festival of textile culture in Haslach (*Textile Kultur Haslach*) which is held every summer for 14 days and which is visited by tourists from Europe and overseas. More than 150 people additionally visits the courses and seminars associated with the festival;

b) day trippers in this area can be totally devoted to textiles;

c) the private initiatives of (older) weavers on teaching traditional weaving techniques find huge acceptance;

d) the brochures of the different local tourist organizations of the region have integrated the *Mühlviertler Weberstrasse* and most of them are even focused on the tradition of textiles and weaving in this region;

e) day trippers are attracted by the coordinated selling of weaving products with gross sale prices in the factories;

f) other European and especially overseas tourists are attracted by the linen products and their design, which are unique and are now marketed in a professional manner. Tourists are willing to pay high prices for products that are not available in the USA or Japan.

The Web Presentation of the Region and the *Mühlviertler Weberstrasse*

The *Mühlviertler Weberstrasse* is presented in the Web with the intention of attracting tourists characterized by their high level of education, cultural interest and environmental consciousness. Tourism in this region is primarily focused on the second or third holiday of European tourists. Secondary targeted groups are overseas tourists and tourists who spent their primary holiday in this region (e.g. to visit some in-depth courses or workshops). This targeted population promises a good return on investment. Recently the Upper Austrian Tourist Board decided to present its tourist information in the Web by using TIScover (www.tiscover.com) because of its potential to integrate the retrieval of tourist information and reservation of tourist services (e.g. hotel rooms, courses and concerts). Another advantage of the Web is the distribution, creation and maintenance of Web pages by tourist providers (e.g. hotels) themselves. Another important characteristic of TIScover is the multilinguality and multi-currency-features which are essential for a tourism web product (see also Tjoa and Werthner, 1996; Burger et al., 1997).

However, the *Mühlviertler Weberstrasse* had organized its presence on the Web even before the initiative of using TIScover for the presentation of Upper Austria. This enables the tourist providers in the region to handle direct reservations via the Web.

Currently the *Mühlviertler Weberstrasse* is described in detail with the following features:

a) the economic value of *Mühlviertler Weberstrasse*;

b) 'adventure and experiences' along the *Mühlviertler Weberstrasse*;

c) the *Textile Kultur Haslach*;

d) the history of Mühlviertel's linen;

e) the production of Mühlviertel's linen;

f) detailed information concerning the different Mühlviertel linen towns.

A similar approach to the *Mühlviertler Weberstrasse* has been taken in another region of Austria in the Waldviertel. Here the *Waldviertler Textilstrasse* – an initiative which even started earlier than the *Mühlviertler Weberstrasse* – has also been proved to be very successful. In this region three textile museums provide the starting points of the route through 40 towns where the tourist can find traditional textile production in manufactures or even home-made textiles and objects of textile art.

The Role of Hand-Made Lamb's Wool Production in Styria's Tourism

Another Austrian example of the role of textile in tourism can be seen in the local initiatives in the region of Styria (*Steiermark*), where the spinning, knitting and weaving of lamb's wool is one of the foci of interest of the local tourist organizations.

In this region wool is woven or knotted into carpets, curtains, table cloths, etc. But also artistically-designed woven paintings and works of art are made of this hand-spun material.

Wool-stuffs are naturally dyed with colours made of plants, roots, etc. and thereafter spun and woven by hand. This wool-stuff is the basic material for traditional and modern dirndl dresses, the traditional costume of the female population of Austria. Special efforts are invested on the details of these dresses (e.g. original designed buttons made of deer-horn or ceramics).

Besides the selling of these local products in souvenir shops and local museums (e.g. in *Kapellen*), courses on dyeing, weaving, and the creative design of carpets and wool paintings are offered to tourists.

The initiatives described above are an important vehicle for a region which has not yet exploited all of its tourism relevant potential (in contrary to other Austrian regions such as the Tyrol or Salzburg). An indication of this fact is that the Web presentation of these regional activities is not yet presented in a professional manner.

The Presentation of Textile Art Exhibitions as a Tourist Attraction

As an example for the attractiveness of textile art exhibitions in museums we would like to mention the exhibition of the Ethnological Museum of Vienna entitled *LebensMuster – Textilien in Indonesien* (*Life Patterns – Textiles in Indonesia,* 14 September 1995 – 29 February 1996), which was one of the most complete exhibitions of Indonesian textiles ever held outside Indonesia. Many tourists, connoisseurs, and experts throughout Europe came to Vienna just to visit this exhibition, which is excellently described in the 255-page catalogue. More than 1,000 textile objects from Indonesia were shown in this exhibition.

Due to the initiative of the Austrian Federal Ministry of Education, arts Web-pages were implemented for all federal museums and exhibitions. The dates of these exhibitions were disseminated through the Web and other international media (e.g. CNN, international magazines, etc.)

Further Potential of Web Presentations of Textiles in Austria

The designs of textiles of famous artists of the *Jugendstil* (Viennese Art Nouveau, e.g. Hofmann) and Bauhaus which are in the possession of famous private Viennese cloth stores (e.g. Backhausen) may have potential for Web presentations to stimulate tourism. Many of these patterns are still being produced. The original design paintings, sketches and specimens are still kept in private archives but not yet presented in the Web. The presentation of these designs in the Web could be a rewarding task for tourism Web pages in the future. This is especially relevant because of the ongoing popularity of the *Jugendstil* (e.g. Klimt, Schiele) and the Bauhaus (e.g. Gropius). The different Austrian experiences clearly show the potential of using (traditional) textile production and textile art as a vehicle for tourism marketing and creation of value in the tourism business. The World-Wide Web is an extremely suitable instrument to promote tourism in this area.

References

Britannica Online (1997), *Encyclopaedia Britannica, <http://www.eb.com:180/cgi-bin/g?DocF=micro/211/57.html>*

Burger, F. et al. (1997), 'TIS@WEB – Database supported Tourist Information on the Web' in *Proceedings of the 4th International Conference on Information and Communication Technologies in Tourism*, Springer Verlag, Vienna and New York.

Heindl B. (1992), *Textil-Landschaft Mühlviertel – Mühlviertler Weberstrasse*, Linz, Edition Sandkorn.

Tjoa, A.M. and Werthner, H. (1996), 'Interfacing WWW with Distributed Databases' in *Proceedings of the 3rd International Conference on Information and Communication Technologies in Tourism*, Springer Verlag, Vienna and New York.

33 The Role of Tartan in the Development of the Image of Scotland

RICHARD BUTLER

Introduction

Batik and tartan are two of the most easily recognizable textiles in the world, but while batik is found in a variety of forms in different parts of the world, tartan tends to be associated very strongly with one country in particular, Scotland. So strong is this association that to many people the sight of tartan automatically triggers the thought of Scotland. To those wishing to sell a product, such an association is a powerful starting point and in the case of Scotland and Scottish tourism, tartan has been used as part of the marketing of Scotland for over a century (Ambrose, 1989). As will be discussed in this paper, however, the use of tartan and the images which it promotes is not without controversy, particularly in modern Scotland, and there exists considerable opposition to continuing to use such images in the promotion of Scotland.

The success of such promotion in the past, however, goes a great deal of the way to explaining how Scotland has managed to establish and maintain a successful tourist industry for the past 200 years. Most successful tourism destinations of the modern era possess a selection of attributes which, although not identical from one location to another, tend to have certain characteristics in common. In general the destinations are readily accessible to major markets, reasonable in cost, have good weather, and possess a variety of activities and facilities of which the visitor can take advantage. They increasingly tend to possess heritage features, either cultural or natural or both, used as supplementary attractions. It is relatively rare for locations to be successful in the long term as tourism destinations without such combinations of features. It is equally uncommon for tourism destinations to be able to maintain their attractiveness over long periods of time unless they have unique and often spectacular features, such as Niagara Falls or the Pyramids, or are key locations

323

such as the capital cities of London or Paris, or transportation hubs such as Singapore and Hong Kong. It is perhaps puzzling, therefore, to imagine Scotland as a successful tourism destination over the long term. Here is a small country, on the periphery of Europe, long plagued by poor access and transportation, costly compared to many peripheral areas, with a climate which is very unattractive to most visitors (and residents) and with limited facilities for tourists compared to other destinations. Yet this country has successfully attracted tourists for over 200 years as a destination, and to a considerable degree seems well placed to continue to attract tourists into the mid- to long-term future.

To understand the paradox which Scotland represents as a tourist destination it is necessary to interpret its origins and the image which it has presented and continues to present to potential visitors. Although this image as a whole is a contrived creation, it has endured for almost two centuries and shows little sign of declining in appeal. Its origins are clear and well documented. An essential element of this image is the use of tartan, the patterned and multicoloured woollen cloth that has come to symbolize a somewhat stereotyped noble savage, in the same way that for a long time the Red Indian symbolized the American West. The use of tartan and related features such as bagpipes, castles, mountains and red deer is not without controversy, and various authors and elements in contemporary Scottish society view such elements with disdain and concern. This paper will suggest, however, that despite some undoubted historical errors of fact and gross distortion of other features, there is more than a measure of validity and appropriateness with the use of tartan as one element in the image of Scotland that still has contemporary value.

The Image of Scotland

Compared to many countries in the world Scotland has a very strong and distinctive image, even if this image is stereotypical and artificial to a great degree. Over the years of teaching students in tourism courses, this author has frequently surveyed the image held by students of selected countries. The image for Scotland has remained consistent and strong for two decades amongst this admittedly unscientific sample. It includes primary features such as mountains, tartan, bagpipes, castle and kilts, and secondary features such as highland dancing, haggis, heather, golf, Balmoral and lochs/lakes. It is, as Gold and Gold (1995) have noted, the 'shortbread' or 'chocolate box' image of Scotland, the pastiche of scenes portrayed on containers of merchandise which are from, or purport to be from or related to, Scotland.

The importance of the overall image of Scotland as a country to the specific image of Scotland as a tourism destination is critical. While the desire of Scots, particularly those charged with marketing their country in the post-industrialized world, is to portray their country as being up-to-date and blessed with modern facilities, particularly in transportation and communications infrastructure, the appeal of Scotland to tourists would appear to be very much tied to an image based on a limited and mostly inaccurate historic picture, very much at odds with the modern scene (Gold and Gold, 1995). In portraying the modern Scotland, therefore, promoters are indirectly at least, in conflict with the very successful tourist image of Scotland, a point which is returned to later.

Tartan in Historical Context

Tartan is essentially a plaid cloth, traditionally made of wool, and in its original form, used only natural colours of wool. The oldest known example of tartan is what is known as the Falkirk sett, and is a scrap of cloth dating from the third century AD, found buried at the Antonine Roman Wall. It had been used as a stopper in an earthernware pot, and is made of undyed brown and white wool from the Soay sheep (Urquhart, 1994, p. 8). The early tartans were believed to be simple checks of two or three colours, obtained from plants, berries, roots and trees which were found locally in the area of production. Over the centuries specific basic patterns produced by local craftsmen became associated with the major family in their area, primarily because those producing the cloth would sell their produce locally, within a relatively small geographical area. Thus most people within an area would wear a similar cloth, and this gradually became recognized as the tartan of the clan or extended family of that region.

Reference is made in the fifteenth and sixteenth centuries to the kings of Scotland wearing tartan, and patterns by this time were beginning to be regarded as proprietary (Collins, 1991). These early tartans were still simple in design, incorporating normally only two or three colours and relatively simple patterns. These patterns gradually acquired the names of the clans or families with which they were associated, and by the end of the seventeenth century were widely recognized and accepted. It is of crucial importance to realize, however, that the use of tartan and the feudal clan system of which it was a part was confined by this time to the north and west of Scotland, that area known as the Highlands and Islands, an area which remained loyal to the Catholic Stewart royal family, which was in exile from Great Britain from 1688, and it was to rise in rebellion against the rest of Britain three times in the first half of the seventeenth century.

The final rebellion, which ended in the Battle of Culloden, the last battle fought on British soil, in 1746, saw the end of the clan system and the introduction of military rule over the Highlands of Scotland (Prebble, 1961). As part of the policy of repression, the bearing of arms and the wearing of the traditional Highland dress, including tartan, were prohibited. The feudal system of the Highlands was broken by force and by imposed social and economic changes and was accompanied through the second half of the eighteenth century by extensive changes in the agricultural patterns of the region and subsequent forced emigration of much of the population. The 'Highland Clearances' (Prebble, 1963), which saw people removed and replaced by sheep, resulted in massive emigration to other parts of Britain, and particularly to Canada, Australia and New Zealand. The bulk of the population which remained was often relocated from the interior valleys (glens) to the coast, and traditional clan chiefs were replaced by absentee landlords, many of whom were from England or southern Scotland, who allowed the properties they acquired to fall into disrepair. They were often replaced in the nineteenth century by pseudo-castles and shooting lodges, many used for the summer periods only, frequently based in design and location on Rhineland castles and built in what is known as the Scottish baronial architectural style. Examples include Carbisdale Castle in Sutherland, the Trossachs Hotel in the central Highlands, and perhaps above all, Balmoral Castle in Royal Deeside, of which more will be said later.

Not surprisingly, given its recent history of armed insurrection, Scotland was not seen as a holiday destination by anyone in the eighteenth century (Youngson, 1974). Perhaps unexpectedly, however, the area did receive a large number of highly significant and noteworthy visitors who can be regarded quite justifiably as the precursors of later more conventional tourists (Butler, 1985). For the most part these visitors fell into three distinct groups; those involved with the military occupation and administration, those engaged in scientific enquiry, and those involved with the world of letters, music and art.

It is the third group of visitors, however, who had by far the greatest effect upon establishing the image of Scotland for tourism and in developing the fledgling tourism industry. This group consisted of authors especially, but also musicians, poets and artists. Indeed, a glance at the list of authors who visited Scotland between 1770 and 1870 reads like a *Who's Who* of English literature, including Defoe, Johnson and Boswell, Burns, Dickens, Coleridge, Wordsworth, Southey, Tennyson, and Scott. What brought them in the later years was essentially the romantic appeal of the area, but that appeal was in itself partly created by these visitors. Defoe (1974), who was one of the first of this literary group to visit this area, toured Scotland in the late eighteenth century as part of his tour of Great Britain, and described a visit to the Highlands as akin to a military expedition, an appropriate term given the military presence there at that time. His writings were some of the first to document the attractiveness of the area, however, and set the tone for other visitors. Johnson and Boswell soon followed and their comments, some delightfully pithy, such as Johnson's on discovering that he had lost his walking stick on the Isle of Skye, when he remarked that this had contributed significantly to the amount of wood on the island, gave widespread publicity to the Highlands (Boswell, 1852). Despite the tribulations their travel involved, their exploits did much to instil interest among prospective visitors to the area.

In overall terms, however, it was the travel and writings of James Macpherson which truly placed highland Scotland and the Gael on the artistic and literary map of Europe in the eighteenth century. Macpherson produced and published what he claimed to be original poems in the epic style, by a Gaelic poet called Ossian (Macpherson, 1765). These poems proved tremendously popular throughout Europe and the image and exploits of the tartan-dressed Gaelic hero Fingal and his cultural and spatial context attracted great curiosity. Ossian was reportedly the favourite poet of many European celebrities from Beethoven to Napoleon (Gold and Gold, 1995). 'The combination of Celtic heroes, highland scenery, chivalry and fine emotions and sensibility was an irresistible combination. It also led to new ways of seeing and representing Scottish landscape' (ibid., p. 54). This interest lasted for several decades and resulted in many visitors from Europe coming to the Highlands, including celebrities such as Mendelssohn, who composed the Hebridean Overture after visiting the western isles in 1820. Fingal's Cave on the island of Staffa is one of the few remaining pieces of evidence of this phenomenon, which disappeared quickly after Macpherson finally admitted to writing the poems himself, destroying the legend he had so effectively created (Gold and Gold, 1995). The image of the Gaelic hero remained, however, and was to subsequently be re-established on an even larger scale by Sir Walter Scott a half century later.

However, the wearing of the Highland dress and tartan was still prohibited and the proscription was not repealed until 1782. For close to half a century the only tartan manufacturing that remained was in the hands of the UK (English) military and suppliers from the south of Scotland. The revival of tartan was encouraged by newly-formed Highland Societies in London in 1778 and Edinburgh in 1780, building on the changing image of the Highlands, and later the military, which made heavy use of the Highland regiments in the late eighteenth and early nineteenth century, in the Napoleonic Wars and those which followed in the nineteenth century, such as the Indian Mutiny, and the Crimean War. By 1822, when George IV visited Edinburgh, there were over 200 tartan setts recorded (Urquhart, 1994). While Robert Burns, Scotland's national poet also visited the Highlands, his travels were confined to a few areas and two visits only. He wrote several poems during and based on these two visits, but shared the Lowland Scots' traditional apathy towards the Highlands and its inhabitants in general, perhaps based on an unfamiliarity with the landscape and customs, and memories of the tartaned 'Highland Host' descending on the Lowlands to replenish their flocks and purses through the power of the sword.

One other group of visitors did exist, and they were becoming more established. These were the absentee land owners and their friends, who had begun to visit their often vast estates in highland Scotland during the summer, and to partake in the traditional highland gentry's activities of hunting and fishing. The hunting of red deer in particular, and the fishing of salmon, became prized attributes of these estates, many of which consisted of large expanses of moorland and mountain, with very little agricultural value, even for the recently-introduced sheep, and whose forestry potential was not yet appreciated. A few journals were written (Brander, 1973; Scrope, 1847), revealing the wealth of such resources and clearly whetted the appetites of sportsmen in the south. Demand began to develop for access to such resources, and the principal of renting and leasing the shooting and fishing rights to some of the estates became established. The first commercial 'let' of shooting property occurred in 1800 (O' Dell and Walton, 1962, p. 332). As Orr (1982) has pointed out, as this practice expanded, clearances of the population again took place, this time for deer and grouse, although on a much more localized and smaller scale than those relating to sheep.

At the end of the eighteenth century, therefore, Scotland was a destination for a small and distinctive tourist market, but was still singularly ill-equipped to receive any visitors. Access to Scotland was poor and within Scotland, especially outside the central lowland belt, it was extremely difficult. 'In the islands there was no roads, nor any marks by which a stranger may find his way' (Johnson, 1775, p. 48). Land transport was poor to nonexistent, except for a few military roads and bridges, rare although considerably improved due to the efforts of Thomas Telford (Haldane, 1962). While the Caledonian Canal bisected Scotland from southwest to northeast, its use was light, and much of the Highlands remained relatively inaccessible. Accommodation for visitors was similarly poor or unavailable, unless one was fortunate enough to have introductions to the landed gentry. Thus tourist travel was akin to the old meaning of travel, 'travail', and those who made the effort normally had specific attractions which they wished to see and more than a passing interest in the region.

Sir Walter Scott and Transformation of the Highlands

As the area gradually lost its threatening reputation as the home of a hostile population, as transportation slowly improved and inns began to be established and as the amenity resources of the area became appreciated (Gilpin, 1789), the scene was set for the total transformation of the image of the region. The individual responsible for this transformation was the novelist and poet Sir Walter Scott, appropriately known as the 'Wizard of the North' for his ability to create imagery (Lockhart, 1906). Scott virtually single-handedly changed the image of the Gael and his homeland from one of despair, unattractiveness, savagery and violence, to one of triumph, beauty, nobleness, and above all, romance. Through his poems and novels he created and developed a mythology akin to that of King Arthur and the Round Table, based on a mixture of reality and artistic licence and presented in a style which perfectly caught the imagination of Victorian society. At the heart of much of Scott's writing was the Highlander, a Gaelic 'noble savage' dressed in traditional kilt and tartan, with a strict code of honour and an extensive cultural history.

Scott's books were best sellers on a scale never witnessed before, and there is extensive evidence that his work had a phenomenal effect upon changing the image of Scotland (Butler, 1985; Gold and Gold, 1985). His greatest single feat was to stage manage the visit of George IV to Edinburgh in 1822, complete with clothing the king and his court in newly-created tartan and pseudo-Highland clothing (Finlay, 1981; Prebble, 1989). This set the stamp of ultimate approval upon the idea of visiting Scotland and wearing 'highland dress' on ceremonial occasions.

Due to the efforts of Sir Walter Scott, the royal seal of approval was added to the now highly fashionable Highland costume by a kilted King George IV. The chiefs of the clans were commanded to attend the King at Holyrood Palace in Edinburgh wearing their Highland dress. This royal patronage was continued and extended by Queen Victoria in her passion for things Scottish (Urquhart, 1994, p. 8).

Scott's success has been discussed at length elsewhere (Butler, 1973; Gold and Gold, 1995) and is unparalleled elsewhere, either in the extent of transformation or in its lasting effect. 'He effectively wrote the script for the promotion of Scottish tourism through the nineteenth and twentieth centuries' (Gold and Gold, 1995, p. 195). The image of the Scot became, and has remained, the image of the tartaned Highlander, despite the fact that Highlanders have always represented a distinct and atypical minority of Scots throughout the history of Scotland (Trevor Roper, 1983).

Scott's efforts were reinforced both deliberately and coincidentally by his peers. Many authors and poets were attracted to Scotland by Scott's works, in particular Wordsworth, Coleridge, Southey, and Dickens. Their own literary efforts also served to stimulate interest in Scotland, and the Highlands in particular (Scott, 1994). The great contemporary landscape artist Turner was engaged to illustrate editions of Scott's works and these and other paintings he produced independently on visits to Scotland, similarly evoked interest in the area. His artistic support of Scotland's image was reinforced by Landseer, Queen Victoria's favourite artist, who painted many Scottish scenes and backdrops for his work, and others (Butler, 1985).

Two other noteworthy features shaped the development of tourism in Scotland and its image in the early to middle nineteenth century, one relating to transportation, the other to royalty. The development of the railway system in Great Britain allowed access to many areas previously not easily visited. While the railway came late to the north and west of Scotland, steam ship services were established early to the west coast and the islands, and served as the basis for much of the tourist travel (Butler, 1973). Real growth in tourism came with the efforts of Thomas Cook, who began organized tours to Scotland from 1846 using rail as far north as possible and then steamer and stagecoach (Cook, 1861). Cook's tours visited many of the sites described in Scott's novels and poems, particularly the Trossachs, the central Highlands and the Isle of Skye. Cook popularized the concept of the 'Tour to the Highlands and Islands of Scotland', in many case following closely the routes of Defoe, Johnson and Boswell and the early pioneers of 'leisure' travel in the area.

The second noteworthy event was the visits by Queen Victoria to the Highlands from 1842 onwards, culminating in the purchase of the Balmoral property on Deeside and the subsequent construction of the present Balmoral Castle in 1855 (Brown, 1995; Duff, 1983). The great fondness of Queen Victoria for things Scottish, including tartan, completed the transformation of the image of the Highlands and greatly popularized the area and elements of its culture such as tartan. 'Prince Albert not only manifested the tartan image by wearing the kilt but also designed and reproduced tartan kilts for the retainers of the royal family' (Jarvie, 1991, p. 68). This popularization of tartan was copied by other aristocratic landowners in Deeside and 'When Queen Victoria came to the area a lot of landowners got the people who worked for them [stalkers, tenants, keepers, etc.] ... rigged out in tartan ... Queen Victoria rigged out all her retainers in the Royal Stewart tartan' (President of the Braemar Royal Highland Society, 4 July 1986, commenting on the impact of Queen Victoria's attachment to the Highland Gathering, cited in Jarvie, 1991, p. 71).

The establishment of a royal summer holiday in the Scottish highlands represented an endorsement of the idea of Scotland as a holiday destination of the highest social order. To the aristocracy a Scottish estate became *de rigeur*, and tartans and tweeds, stalking rifles and salmon rods and a Scottish 'season' became a part of the social order in Victorian society. To those who could not aspire to such a level of investment of money and time, Cook's tours and individual holidays represented an acceptable alternative. By the 1850s Scotland was firmly established as a respectable holiday destination by the highest social groups in the land and the wearing of tartan in that setting and on ceremonial occasions was eminently fashionable.

Tartan in the Contemporary Scene

The increasing popularity of tartan through the nineteenth century, stimulated as noted above by Queen Victoria and Balmoral, has continued through the twentieth century. New tartans, many unrecognizable to ancient or modern Scots alike, now grace the human form and many other items besides. Tartan has become used as a form of decoration and embellishment for many items from books to luggage, and from linen to food and beverages. It has come to symbolize Scotland and things Scottish, and is often used instead of, as well as with, the name of Scotland. It was almost inevitable in these circumstances that it became used for promoting the image of Scotland in the context of tourism. On one hand this increasingly widespread use of tartan has received support:

Tartans have always formed part of Scotland's historic heritage and it is a compliment to their country of origin that they have become so widespread throughout the English and Gaelic speaking world. They are probably more popular now than they have ever been because they have come to symbolize the spirit of families, clans and district, and more recently, corporate institutions (Teall, 1994, p. 6).

On the other hand, the recreation of tartans and associated myths has given rise to many criticisms. Caughie (1989, p. 93) comments ' From popular, as well as legitimate cultural representations, three immediately recognizable Scottish "myths" seem to emerge .... Tartanry, the most obvious, the most notorious, and in its tourist manifestations, the most embarrassing ...', and adds 'the tartan monster is still alive and roaming the land (possibly disguised in the new tartan uniforms of the custodians of Heritage Building)' (ibid., p. 92). Other modern writers link tartan with other images of Scotland and castigate them with the same brush: 'It's not enough to get rid of the scotch myths of haggis, tartan, whisky and Harry Lauder' (Clark, 1989, p. 9).

The concern of many contemporary writers is that tartan and the associated images of the Highlander are inappropriate representations of Scotland as a whole at any time and very much less so at the end of the twentieth century. They refer to the recreation of tartans and the appearance of 'tartany' as a 'standard kitsch symbol of Scottish cultural identity' (Nairn, 1981, p. 91). Yet, as Jarvie (1991, pp. 25–6) argues,

A number of modern historians have tended to argue that the dress of the Highlander, in particular the notion of tartanry, was very much an invented tradition which emerged during the nineteenth century (Trevor Roper 1983). Yet there is a great deal of evidence which suggests that while the production of clan or family tartans did not exist before the nineteenth century on such a large scale as it does to day, it is in fact wrong to assert that the tartan tradition of the Highlanders was an 'invented tradition'. Yet it is important not to divorce the meaning of tartanry in all its various forms from the original context in which it was developed. Furthermore, it is important not to consider the wearing of tartan today as a completely novel development divorced from earlier traditions ... Within its original social context a tartan ... gave expression not just to a form of dress but to a whole way of life.

It is the transformation from tartan as a material with some spatial and family ties within a specific part of Scotland some two centuries ago to tartan as decoration and form of embellishment of many unrelated items and as a symbol of a whole nation that raises the ire of many contemporary commentators. Tartan has become synonymous not only with a part of Scottish history, but with many, if not to some people, all aspects of Scotland today.

The third factor which, in a multitude of different forms, continues not only to define a Scottish cultural identity but also Scottish sporting identity, is tartanry. Both the view of the emigre and modern kailyardism have relied heavily upon tartany as the dominant expression of Scottish culture. Tartan football armies and Highland Gatherings, and Games, immediately spring to mind. Highland Games programmes and marketing brochures, both at the level of local games and the level of the Scottish Tourist Board are heavily dependent upon tartan imagery .... Such romantic symbols, upon which Scotland has become culturally and to a degree economically dependent, are quite divorced from historically lived experience. It is doubtful if the Highlanders of the pre-1745 epoch envisaged their ultimate inheritance as being that of pictures on whisky bottles, shortbread tins, Highland Games programmes, Scottish tourist brochures or even plaited socks. As an aspect of Scottish culture in general, and of Highland culture in particular, tartanry in its original concept was virtually destroyed towards the end of the eighteenth century (ibid., p. 92).

Why then, we might ask, has tartan endured and in fact become stronger as a symbol of things Scottish and such a powerful signal of Scotland? One answer, I would argue, stems from the old Highland origins of tartan. From its earliest time, it represented the inhabitants of the north and west of Scotland, the Gaelic-speaking clans, who had a distinct and strong culture and image. This is supported by Jarvie (1991, p. 26) who comments 'Certainly ... the wearing of Highland dress and various antecedent forms of tartan help to distinguish the Highlander from the Lowlander'. To many non-Scots the bulk of the Scottish population is visually indistinguishable from most other Caucasians, and much of lowland Scotland is not highly attractive or distinctive as a tourist destination (except perhaps to golfers and lovers of Victorian architecture). What makes Scotland and the Scots distinctive are the symbols of Scottishness, contrived, recreated and only partly accurate though they may be. There is or was no other race quite like the Highlanders of the fifteenth through seventeenth centuries, and their dress and their instrument (the bagpipe) have become instantly recognizable as Scottish. The fact that they never represented all of Scotland is immaterial to most contemporary potential tourists and possibly to many Scots in the Scottish diaspora.

The use of tartan has become firmly established with many Scottish cultural activities carried on in exile, particularly Highland Games and St Andrews and Burns suppers, with such events in North American and Australasia being particularly numerous. Chapman (1979, p. 3) comments with considerable accuracy that 'Scottish culture has drawn heavily on a Gaelic version of cultural identity', and even among critics of tartan and its related paraphernalia there is the recognition that one is dealing with something a little more complicated than a totally artificial construct.

> Almost invisible behind the tartan monster which stalks airport terminals, High Street souvenir shops and Hollywood movies, are real mythic confrontations between progress and tradition, nature and culture, rationality and romance. While many of us would stand back quite happily ... and watch the tartan babies being thrown out with the bath water, there is something in that powerful confrontation, in the complicated appeal of land and landscape, which seems written into Scottish culture and which needs to be recognized and understood (Caughie, 1989, p. 93).

The confrontations which Caughie refers to are still being debated and the recent vote on devolution is powerful evidence of Scotland's desire to see its distinctiveness maintained. In terms of attracting tourists, such distinctiveness is of critical importance and tartan can be a key element in the promotion of Scotland.

Conclusion

In the last quarter century, the overall attractiveness of Scotland as a destination has increased significantly, at a time when competition among all destinations has also increased. Given what has been described of the Scottish image earlier, this may also appear paradoxical, but the reason lies in two main points. The first is that the cultural and natural heritage of Scotland is still perceived of by tourists as genuine and therefore authentic and attractive at a time when much of the tourist world's heritage is becoming obviously artificial and staged. This point is discussed at more length below. The second reason is that the elements that make Scotland attractive as a contemporary tourist destination have increased in quality and variety. While the 'shortbread tin' image is still fundamental, there is a much broader and more realistic base to its appeal.

Scotland has to exist as a specific destination on the tourist stage, it is too peripheral to draw tourists en route to other destinations. That it has managed to remain attractive to tourists for the last 200 years is rather remarkable, given the relative paucity of its attractions compared to many other destinations. Over the past decade the number of visitors has remained relatively consistent

(Scottish Tourist Board, 1995). The key question is whether it will be able to maintain this attractiveness in the future, and a great deal depends on how it is marketed and managed and how the attractions and heritage are maintained so as to keep their integrity and appeal.

Undoubtedly the Scots themselves have a great deal to do with the image of Scotland being so strong. The widely-flung Scottish diaspora represents a recognizable group that often preserves its Scottishness at a level far higher than is maintained in Scotland. There are more active and more numerous Burns' Clubs, St Andrew's Clubs, and Highland Games in Canada, USA, Australia and New Zealand than in Scotland. The image of Scotland is maintained on the world stage also in no small measure through its sports teams, which although perhaps not the most successful, remind the world of the Scottish presence in soccer and rugby particularly. Scotland as a national presence, therefore, gains recognition and image endorsement globally, even if that image is often the tartaned Highlander with bagpipes of Sir Walter Scott. While this author does not disagree with the conclusions of Gold and Gold (1995, p. 202) who state that 'conventional promotions and policy propagate a conservative and incomplete picture of Scotland ... [and] may now limit the potential of tourism rather than expland it', there is a real danger that to abandon the stereotyped image of Scotland would also limit tourism.

It is correct to argue that if all that is marketed is 'Scott's Scotland', such a policy would fail to capitalize on all of Scotland's assets. Missed opportunities already abound, the lack of emphasis and protection for industrial heritage resources is one example, but the fact remains that the tartan image is a very strong and positive one which has been successful for two centuries. To discredit this image or to significantly downplay it because it may be politically incorrect or incomplete as an image for the country as whole could be a mistake of Ossianic proportions itself. The tartan image is uniquely associated with Scotland and highly recognizable. Any alternatives, while perhaps more contemporary and accurate, are less likely to be as popular and successful over such a long period. Combining contemporary elements and attributes with the established image would appear to be the most realistic option in marketing Scotland as a destination in future. The continuing revamping of one brand of Scotch's famous advertising slogan 'Born 1820, still going strong' would seem an appropriate analogy here. In recent years Scotland has benefited, as did Australia, from successful movies and television productions (Riley, 1994).

Just as Australia had *Crocodile Dundee* and *The Man from Snowy River* as well as *Neighbours*, so Scotland has had *Braveheart*, *Rob Roy* and many television series which have been syndicated to much of the English-language television-watching world. Thus the Scotland as portrayed in celluloid and satellite may build on and perhaps modify the image of Scotland which has been conveyed through paper and print for the last two centuries. Potential visitors to Scotland surveyed for their thoughts on marketing images in 1994 confirmed the power of the established image.

> Many respondents thought that a Scottish logo should incoroporate tartan and liked the 'Piper' because they saw this both as a symbol of culture and something uniquely Scottish ... The remaining ideas, although regarded as potentially attractive symbols, were not linked closely enough with Scotland (Cragg Ross and Dawson, 1994, p. 18).

Old images are hard to shake and just as salmon, venison, shortbread and whisky have been gastronomic Scottish favourites for many years, it is likely that tartan (along with bagpipes and the Highlander in native dress), will remain mental and aesthetic favourites for some time to come.

References

Ambrose, T. (ed.) (1989), *Presenting Scotland's Story*, Edinburgh, Her Majesty's Stationery Office.

Boswell, J. (1852), *Journal of a Tour to the Hebrides with Samual Johnson*, London, National Illustrated Library.

Brander, M. (1973), *A Hurt Around the Highlands*, London, The Standfast Press.

Brown, I. (1955), *Balmoral. The History of a Home*, Glasgow, Collins.

Butler, R.W. (1973), *The Tourist Industry of the Highlands and Islands*, PhD thesis, Glasgow, University of Glasgow.

Butler, R.W. (1985), 'Evolution of Tourism in the Scottish Highlands', *Annals of Tourism Research*, 12 (2), pp. 371–91.

Caughie, J. (1989), 'De-picting Scotland: Film, Myth and Scotland's Story' in Ambrose, T. (ed.), *Presenting Scotland's Story*, Edinburgh, Her Majesty's Stationery Office.

Chapman, M. (1979), *The Gaelic Vision in Scottish Culture*, London, Croom Helm.

Clark, T. (1989), 'Introduction' in Ambrose, T. (ed.), *Presenting Scotland's Story*, Edinburgh, Her Majesty's Stationery Office.

Collins (1991), *Clans and Tartans*, London, Harper Collins.

Cook, T. (1861), *Cook's Scottish Tourist Official Directory*, Leicester, T. Cook.

Defoe, D. (1974), *A Tour Through the Whole Island of Great Britain*, London: J.M. Dent.

Duff, D. (ed.) (1983), *Queen Victoria's Highland Journals*, Exeter, Web and Bower.

Finley, G. (1981), *Turner and George the Fourth in Edinburgh in 1822*, London, Tate Gallery and Edinburgh University Press.

Gilpin, W. (1789), *Observations relative chiefly to picturesque beauty made in the year 1776 on several parts of Great Britain: particularly the High-Land of Scotland*, Volume 1, London.

Gold, J.R. and Gold, M.M. (1995), *Imagining Scotland*, Aldershot, Gower Press.

Haldane, A.R. (1962), *New ways through the glens*, London, Thomas and Nelson.

Jarvie, G. (1991), *Highland Games, the Making of the Myth*, Edinburgh, Edinburgh University Press.

Lockhart, J.G. (1906), *The Life of Sir Walter Scott*, London, Dent.

Macpherson, J. (1996), *The Poems of Ossian: and related works*, edited by Howard Gaskill with an introduction by Fiona Stafford, Edinburgh, Edinburgh University Press.

Nairn, T. (1981), *The Break-up of Britain*, London, Verso/NLB.

O'Dell, A. and Walton, K. (1962), *The Highlands and Islands of Scotland*, London, Methuen.

Orr, W. (1982), *Deer forests, landlords and crofters: the Western Highlands in Victorian and Edwardian times*, Edinburgh, John Donald.

Prebble, J. (1961), *Culloden*, London, Secker and Warburg.

Prebble, J. (1963), *The Highland Clearances*, London, Secker and Warburg.

Prebble, J. (1989), *The King's Jaunt: George IV in Scotland 1822. 'One and twenty draft days'*, London, Fontana.

Riley, R.W. (1994), 'Movie-induced tourism' in Seaton, A.V. (ed.), *Tourism-State of the Art*, Chichester, Wiley.

Scott, P.H. (1994), 'The image of Scotland in literature' in Fladmark, J.M. (ed.), *Cultural Tourism*, Wimbledon, Donhead.

Scottish Tourist Board (1995), *Tourism in Scotland 1994*, Edinburgh, Scottish Tourist Board.

Scrope, W. (1847), *Days of Deer Stalking in the Forest of Atholl*, London, J. Murray.

Teal, G. (1994), 'Introduction' in Urquhart, B. (ed.), *Tartans*, London, Apple.

Trevor Roper, H. (1983), 'The Invention of Tradition: the Highland tradition of Scotland' in Hobsbawm, E. and Ranger, T. (eds,) *The Invention of Tradition*, Cambridge, Cambridge University Press.

Urquhart, B. (1994), *Tartans*, London, Apple.

34 Traditions, Tourism and Textiles: Creativity at the Cutting Edge

NELSON H.H. GRABURN

Introduction

My own research interests are in tourism and in the cross-cultural study of the arts. I have focused particularly those arts created by members of one society which are bought, collected and used by members of another, often called Tourist Arts, and on the dynamic changes of the arts within societies undergoing change, often called Ethnic Arts. Though the subject of this chapter is tourists, textiles and fashion, I shall take a larger view by looking at textiles as markers of high status, identity, and fashion, and at the relation between textiles, clothing and other art forms. Most of my examples are drawn from my research in North America.

Sculpture and Silk Scarves: Indirect Tourism and Canadian Inuit

The first case I wish to talk about is that of the Canadian Eskimo – who call themselves Inuit. In summer 1959 and in many years since I have lived with the Inuit in Northern Canada and watched them creating wonderful sculptures in steatite (soapstone), and making block prints, not for themselves but for sale to outsiders, to the people of southern Canada, the United States and the rest of the world. The Inuit were brought into the world economic system as trappers of white fox furs which had become fashionable as ladies wear early in this century. The Inuit used their earnings to buy new imported materials for the elaboration and decoration of their own skin clothing. When the prices paid for fox furs decreased after World War II, they needed to sustain their monetary income from other sources.

These 'export' arts were quite different from anything they traditionally

Illustration 34.1   Traditional Canadian Inuit women's caribou skin parka, modified with beadwork and colourful cloth panels, made in the 1930s (photograph by author)

made for themselves; the new arts expressed both what they thought the market wanted from their arts as well as what they wanted outsiders to think about them. So we can say that the functions of these very successful cross-cultural arts are:

1   to provide the buyers with elegant arts and crafts;

2   to control the outside world's image of themselves;

3   to make money in an increasingly commercial world.

In the 1960s and 1970s years the Inuit of a number of villages started to make textiles, experimenting with media such as silk screen (serigraphs) which, like block printing, were entirely foreign to them. These creative new textile arts had the same functions of making money, expressing the artists' aesthetic culture, and pleasing the buyers hoping to enter the realm of fashion. Inuit textiles never became as commercially successful as the sculptures and prints. In Baker Lake and other villages they have learned to embroider cloth to make beautiful illustrative wall hangings, not only for sale, but to serve as murals in their schools, churches and community halls; in Pangnirtung they have learned the more complex textile arts of loom weaving, but with slow output and high prices. In Puvirnituk, they even experimented with *batik*, but the Inuit crafts people took so long to make each one that they were commercially unsuccessful. The artists of Cape Dorset are wonderful designers and they have licensed their designs to be used on silk scarves, which have become moderately successful as fashionable women's wear in Canada.

In the case of the Inuit, because of the remoteness and expense of getting there, very few tourists can visit their land, so the arts are exported out of the Arctic. This has been called 'indirect tourism' because the consumers – of the sculptures, prints, and scarves – are experiencing these ethnic arts at a distance, without being able to enjoy first hand contact with the Inuit people.

Noble Crests and New Arts: Northwest Coast Indians

A second case I will consider is the artistic production of the Indians – Native Americans – of the northwest coast of the United States and Canada. These peoples have always lived in an area rich in land and marine resources, which supported a ranked society consisting of hereditary nobles, commoners and

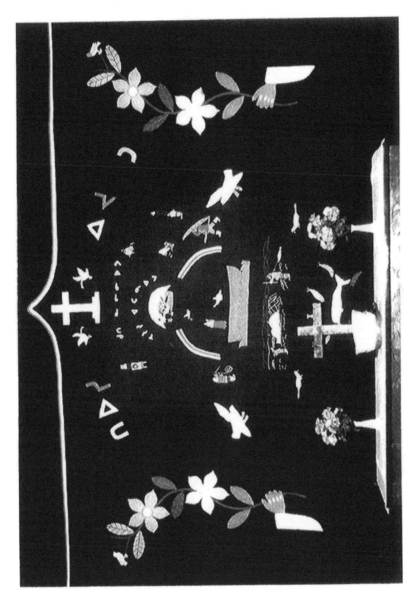

Illustration 34.2   **Canadian Inuit hand-embroidered appliqué cloth, hanging behind altar, Salluit, Quebec, 1970s. Writing in Inuit syllabic script reads: 'Guti nalliniuvut, Jesusi ataniuvit [God who loves us, Jesus our king]' (photograph by author)**

Illustration 34.3   **Experimental batik, made by the Inuit of Povirnituk, Quebec, 1968. On the left, Mlle Ouilette, the batik instructor for the Inuit Cooperative; on the right, Mrs Vola Furneaux, wife of the Canadian government administrator (photograph by author)**

Illustration 34.4   Canadian silk scarf, using Inuit design under license from the West Baffin Eskimo Cooperative, 1980s (photograph by author)

servants. They were prolific artists who enjoyed an elaborate ceremonial life in their plentiful leisure time. These people, known as the Kwakiutl, Haida, Tsimshian and Tlingit, long ago developed striking formal two-dimensional art styles which displayed their historical mythology and their crests which upheld their inherited titles and noble rank. These were carved on totem poles, painted on houses and woven into their clothing.

In the nineteenth century, they were absorbed into the new nations of Canada and the United States and they became rich in trade goods by selling their abundant fish for export. They intensified their ceremonies, created new arts forms with imported materials, such as cloth, beads and buttons, but they suffered badly from European diseases and from Christian missionary attempts to stop the native religion. They started making expensive souvenirs of black argillite slate for rich visitors, and later made cheaper souvenirs for the mass tourists.

In the 1950s and 1960s, they underwent a social and artistic revival. Missionary pressures ended and civil rights were extended to minority peoples in North America and the rest of the world, through de-colonization. They have revived their artistic apprenticeship system, and their ceremonial life. They, like the Eskimos mentioned above, developed new art forms with new materials and techniques, such as serigraphs (silk screen) on paper and textiles. And some have also tried to adapt their brilliant high status clothing for the world's fashion wear market.

Their homeland, not quite as remote as the Canadian Inuit in the Arctic, is somewhat accessible to tourists. Tourists buy both cheap souvenirs and fine arts, and are particularly attracted by the symbols of Indian identity. But the majority of the contemporary arts are brought for sale in the city art galleries, and are exported far and wide. So here we have an example of mixed direct and indirect tourism.

Textiles and Fine Arts: Status, Identity, and Tourism

Fine textile arts have always been markers of high status, and have played a large role in the ceremonial life of traditional peoples. Long before the world of modern tourists, explorers and travellers marvelled at the exquisite productions of other peoples and have endeavoured to collect them. This is true of archaeological arts, such as the feather cloaks of the nobles of the Inca and other Peruvian peoples. Such rare objects which used to be high status clothing of the Inca now form high value artefacts in Western museum

Illustration 34.5   Kwagiulth dancers wearing woven Chilkat blankets, Hagwilget, British Columbia, 1983 (photograph by author)

collections. Similarly, the feather cloaks collected in Hawaii at the time of first contact, full of *mana* that is the tabooed spiritual power of Hawaiian royalty, have undergone the same transformations – destruction if left in situ, or collection by elite museums. Tourists may flock to see these archaeological and historical examples in the museums of Europe and the Americas, which in turn may prompt them to go as tourists to the places where these are originated.

Other fine art and textile traditions have not been entirely separated from the people who produced them. In the case of the Maori of New Zealand, the prestige textiles may be more than a hundred years old and still remain in the hands of the Maori nobility for whom they form part of their material and symbolic heritage. The same is true for the valuable ceremonial 'gift' hats of the Native Californian Pomo Indians, and the crests of the northwest coast people mentioned above. Luckily, there are still many societies who have not lost their traditions. Like the Fijian women who wear barkcloth *tapa* clothing for their wedding ceremonies, people can still make and use their high status material culture such as clothing in rituals which are still extant. In some post-colonial countries, the fine textile traditions of one or more of the formerly tribal peoples have been adopted as national heritage. In Ghana in West Africa, the brilliant *kente* cloth of the Akan-speaking Ashanti peoples has become the national dress for the new ruling classes of all backgrounds. It is used on formal occasions within the country and it is worn with pride by Ghanaians travelling abroad. This is very similar to the present status of Javanese batik for Indonesia.

These examples of high status ethnic arts or sumptuary goods are powerfully attractive to tourists, and they have become important for the world of tourism in a number of ways. (1) they may come to symbolize the identity of the people who make them who themselves become tourist attractions, (2) they may become souvenirs or collectables ranging from cheap imitations to expensive fine arts, and (3) the striking forms of objects, designs and motifs may be appropriated by modern nations as logos or icons in the competition for visibility in the world of tourism. Let me give a few examples.

The Navaho Indians of the southwestern United States are well known for their textiles. The Navajo learned to weave about 130 years ago. For over 100 years they have exported their woollen rugs to consumers in the cities of North America, a form of indirect tourism; at the same time their traditional women's dress fashion and the women's weaving on their looms has long attracted tourists to their land. The Navajo also learned to make silver jewellery and nowadays both rugs and jewellery are sold as souvenirs to tourists on the

Illustration 34.6   Modern Ashanti *Kente* cloth, Ghana (photograph by author)

spot and exported to stores all over the world. Those who come to the American southwest for ethnic tourism are also attracted to the traditional ceremonial life of the nearby Pueblo Indians such as the Hopi, and models of their costumed Kachina god dancers are the major form of souvenirs produced for tourists. During the 1940s and 1950s, Frederic Douglas, Curator of the Department of Native Art in the Denver Art Museum, tried to bring Native American clothing into use among mainstream Americans by means of a travelling 'Indian Fashion Show' in which women college students modelled modified forms of traditional textile and deerskin garments for non-Native audiences (Parezo, 1999).

For centuries the Kuna Indians of Panama have been visited by foreign explorers and sailors, but they still retain some autonomy within Panama. They learned to make striking *mola* blouses when they acquired cloth, scissors and thread about 90 years ago. These textiles have become the uniform of Kuna women, and serve both as a marker of identity for the inhabitants of different islands and as expressions of the women's artistic skills. In addition, the creativity of the women constantly leads to exciting innovations, and one can trace generations of Kuna fashions in collections of these textiles.

In the past 40 years the Kuna homeland, an archipelago near the Atlantic end of the Panama Canal, has been visited annually by thousands of tourists on cruise ships. The tourists not only want to see the colourful Kuna, but they buy the used *molas* as souvenirs. Back in the tourists' home countries many people do not understand that these textiles are clothing, and they mount the embroidered front and back panels on frames hung like pictures!

In Mexico, the textiles and clothing of many Indian groups are attractive to tourists. Not only do modern middle class tourists want to buy these crafts as souvenirs but, since the 1960s many tourists want to wear them. So the indigenous Mexicans have to modify their *huipil* shirts to accommodate the very large arms and bodies of Caucasian tourists. I am sure the same is true in Indonesia!

All modern nations have to create and maintain distinctive identities both for the benefit of their own inhabitants, and for visibility in attracting international tourists. One problem for many post-colonial nations has been the choice visible icons of identity which distinguish the new nations both from their former mother countries, and from other ex-colonies of the same countries. For instance the countries of Latin America were all colonies of Spain or Portugal, and New Zealand, Australia and Canada all used to be English colonies.

In their adoption of national and touristic icons, countries may draw upon

Illustration 34.7 **Kuna women in traditional clothing with reserve appliqué Mola blouse, Panama, 1996 (photograph by permission of Prof. Mari Lyn Salvador)**

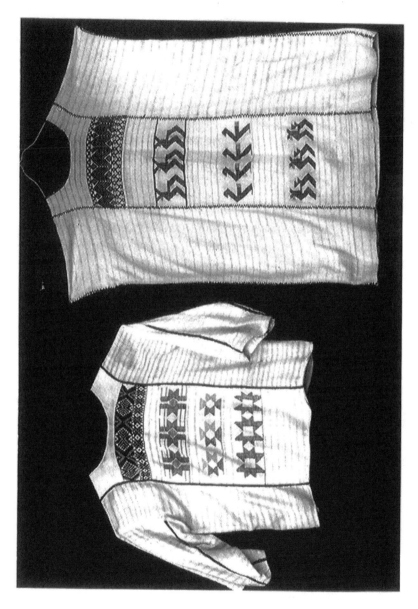

Illustration 34.8 **Mexican Indian cotton Huipil from Oaxaca in two forms: original sleeveless form for native use and modified form for sale to *gringa* tourists, with enlarged armholes and sleeves made from the original waistband (photograph by author)**

whatever is distinctive in their natural phenomena or their cultural products and traditions. In many cases the modern countries have appropriated distinctive items of the very people they have conquered or absorbed. For instance, the Canadian government frequently gives Inuit sculptures to distinguished guests and offers them for sale to tourists, while decorating its airports and worlds fair pavilions with large Inuit *inuksuk* sculptures. In this way it can distinguish itself from the United States whose Alaskan Eskimo arts are quite different. However, at recent world fairs, both the United States and Canada have been known to mark their national pavilions with northwest coast totem poles! Australia and New Zealand have similarly drawn upon the designs of their Aboriginal and Maori peoples respectively for distinguishing their aircraft and buildings. Even Japan, an old nation developing a new identity, offers the arts of the Ainu minority people of Hokkaido to both domestic and overseas tourists, and proudly displays Ainu textiles and their designs as part of their national and touristic heritage, along with their better known traditional *kimonos*.

Conclusions

The past century of colonial encounters, and that past half century of post-colonial national building and transnationalism have witnessed an accelerating rate of production, change and circulation within the world art system. Both geographical barriers and hierarchical social strata are being dissolved by the movements of people, and of ideas and goods. Even the most remote or formerly powerless peoples are becoming aware of the world's artistic creations and fashions even if they cannot afford them. At the same time the central powers of the world system are collecting arts and crafts and expropriating motifs and ideas for their own consumption fashions as well and for cheap reproduction and further mass distribution. This clash of formerly distant peoples and cultures has extinguished some traditions but has also fostered exciting hybrid forms and has stimulated new waves of innovation and creativity. Among these artistic products, textiles particularly in the form of clothing, which can be so easily, carried, worn and displayed, are in the forefront of these travelling arts which are making their way all over the world.

Illustration 34.9   Selling contemporary textiles to tourists, Sumba, Indonesia, 1996 (photograph by permission of Dr Jill Forshee)

References

Graburn, N.H.H. (ed.) (1976), *Ethnic and Tourist Arts: Cultural Expressions from the Fourth World*, Berkeley, University of California Press.

Graburn, N.H.H. (1984), 'The Evolution of Tourist Arts', *Annals of Tourism Research*, 11, 3, pp. 393–419.

Parezo, N.J. (1999), 'The Indian Fashion Show' in R.P. Phillips and C.B. Steiner (eds), *Unpacking Culture: Art and Commodity in Colonial and Postcolonial Worlds*, Berkeley, University of California Press.

Conclusions

MICHAEL HITCHCOCK AND WIENDU NURYANTI

One of the main themes running through this volume is how craft makers should respond to technological change. Of concern here is the evolution of a division of labour which connects ever increasing circles of people in commodity production and exchange (Hart, 1982, p. 38). What is significant about this process of commoditization with regard to batik is that newer methods have not entirely ousted older ones. The reasons are not clear cut and may be linked to batik's cultural significance on one hand, and the elegant simplicity of the basic technology on the other. Irwan Tirta points out, for example, that *cap* methods were introduced for purely commercial reasons, though they took time to gain acceptance and never entirely replaced hand drawn batik. A similar observation is made by Ardiyanto Pranata with regard to the north coast of Java where artistic possibilities were enhanced through the availability of imported synthetic dyes. The fact that concerns about batik's future are raised from time to time, paradoxically seems to indicate its resilience. Teruo Sekimoto warns that the tendency to romanticize batik on account of its age should not obscure the vital role it plays in Indonesian development. As Nazlina Shaari argues, technology presents a challenge, not least with regard to computer aided batik waxing processes. Her remit is to enhance this fabric's appeal in a time of rapid technological change while retaining Malaysian batik's distinctive historical and cultural references. Batik's range may be extended with the use of new materials and Pleasance discusses possible applications with regard to fine wool, a prestige fibre sought by international fashion. Sometimes minimal contact with the source of a craft tradition can stimulate highly varied results, a good example being Farkas's use in Argentina of wax resists on paper.

Many of the contributors address the question of quality and the linked concerns of intellectual property rights. This is a particularly pressing issue for the craft community because the advent of tourism and the mass media has facilitated access on a hitherto unprecedented scale. Concerns are expressed about that harbinger of globalization, the appropriation of one culture by another. Appropriation is a double edged sword because it has the potential to raise a craft's profile and prestige, but, if handled insensitively, can cause offence. Some argue in favour of some kind of regulatory body to maintain standards and secure prices. The drafters of the batik declaration at the conference in Yogyakarta had similar views in mind when they referred to batik has being analogous to traditional medicine. Morrow, however, points out that intellectual property is principally concerned with private property rights and that in practice 'cultural property' is difficult to establish because of ambiguities concerning ownership and proprietary rights. The lack of a common technical definition due to the different cultural traditions of batik makers also makes enforcement difficult. The situation is further complicated by the different legal systems of the countries involved in trading handicrafts. Indonesia's, for example, is based on Dutch civil law, whereas Australia's is founded on common law. Morrow suggests that the protection offered by copyright is potentially of more value to batik makers than a system of registered designs. In order to support claims to ownership batik makers should retain original copies of designs Irwan Tirta suggests introducing a batik mark of quality rather like the international woolmark. Morrow argues that this may have particular applications in tourism to enhance recognition of craft products among those unfamiliar with the genre.

Many of the contributors are concerned with maintaining batik's image in the marketplace. Trefois, for example, is concerned about over familiarity with batik as product and the attendant risk of vulgarization. Prices often have to drop in order to secure markets, but often at the risk of downgrading products in the eyes of purchasers, particularly tourists (see Amri Yahya). Underpinning this discussion are questions about what makes batik distinctive.

Several authors draw attention to the need to raise awareness about batik in order to maintain standards. The fact that batik can be distinguished from other textile techniques through the appearance of the pattern on both faces of the cloth enhances recognition.

Many authors, however, refer to creativity of cultural interchange and draw attention to the splendour of the mixed heritage of batik such as that of the Chinese and Indo-Europeans of Java's north coast at the turn of the century. As Ardiyanto observes, this process of cross-fertilization has continued well into the twentieth century. Popularly inspired designs are also encountered, and, as can be seen from Gittinger's account, access to a comic book in the 1930s opened up a whole range of artistic possibilities. In 1969, for example, Ibu Sud absorbed Thai ideas and introduced a variant of Thai dress made of batik. By the late twentieth century a great deal of batik in Indonesia had become less Javanese, more Indonesian and increasingly international.

Much emphasis is placed by the contributors to this volume on the dynamic character of batik design. As Heringa notes, the status of designs may also alter in accordance with broader sociopolitical changes, as has been the case with *batik-lurik*. The link between the rejuvenation of batik and national and regional consciousness is also considered. New nations often choose ancient symbols to emphasize continuity with an exemplary past, though they have to be adapted and reinterpreted to suit contemporary requirements. Woro Aryanti writes, for example, about the choice of Garuda as the symbol of the Indonesia; this man-eagle's successful efforts in freeing his mother from suffering symbolizes the struggle against colonialism in the post independence republic. Nazlina Shaari, for example, notes the use of a batik symbol on the tail fin of Malaysia's national airline. Siva Obeysekere discusses the revival of batik in Sri Lanka in relation to cultural and national esteem. The adoption of long-sleeved batik as formal wear under Ali Sadikin's governorship of Jakarta simultaneously boosted batik production and local pride.

A related theme is the link between textiles and identity and the role of dress in defining not only who one is, but who one is not. Crafts specialists are, for example, mentioned in Old Balinese sources and, as Wayan Ardika argues, rulers have long sponsored craft production to boost their prestige. There remain, for example, fine distinctions between Yogya and Solo batik; and, according to Hertini Adiwoso, it is not just the design but how the garment is worn that is significant. Local expressions of identity, the antithesis of globalization, is often communicated with the aid of decorative fabrics. As Hayman notes with regard to Central America, the adoption of particular kinds of dress may convey political messages and announces the fact that the wearer subscribes to the values of his or her community. Expressions of indigenous culture, though often denigrated by elites, may paradoxically stimulate the interest of tourists. Graburn likens batik to other kinds of what he refers to as 'pride' cloths such as Kente cloth from Ghana, Maori cloaks, Fijian *tapa* and Navaho blankets. In a similar vein, Butler considers the reinterpretation of Scotland's identity with regard to tartan, tourism and national consciousness.

Batik designers are constantly exploring the full potential of this craft but what should not be overlooked is the continuity. Many new developments would not be possible without the underpinning provided by the conservators of batik's heritage We get a sense of this from Itie van Hout's analysis of batiks at the Tropenmuseum in Amsterdam. Puspitsari Wibisono, for example, describes museums as cultural memory banks and discusses the role of exhibitions in spreading goodwill and stimulating interest. Without the extraordinary dedication of the staff at the Rotterdam Museum our knowledge of the history of batik would be considerably diminished. Exposure to ultra violet light, insect damage and handling all contribute to the deterioration of cloth and thus the conservation skills developed in museums are vital in preserving batik as a cultural resource.

What is significant about this volume is the treatment given to batik outside the Asian region. The sinuous motifs associated with art nouveau and art deco may be partly attributable to Javanese inspiration. Some similarity may be coincidental, but Wronska-Friend is able to document specific examples, especially with regard to the work of the Belgium artists Van de Velde. Russian art was also historically influenced by contact with Asia, and this appears to have facilitated the acceptance of batik among certain artists. Blank argues that although these artists have been open to diverse influences, their narrative and mythical subject matter has much in common with Asian traditions. There also exist batik traditions that were not necessarily influenced by Asia, though parallels may be drawn with the Asian experience. Haake and Hani Winotososastro discuss the revived custom of batik egg production in Europe and the apparent anomaly of old customs flourishing in industrial societies. Interestingly, the protective power of batik patterns is encountered in the European context.

What is also significant in this context is the use of batik as a leisure or hobby activity. Interest in batik in Germany, for example, stimulates interest in Java and a desire to visit the source of this craft. Hobby batik reached a high point in Germany between 1970 and 1990 and since then enthusiasm has dwindled. According to Irene Romeo batik is too labour intensive for hobbyists. Silk painting, a related and comparatively easier craft, is beginning to replace batik. European interest in batik in the late twentieth century may be linked to the hippy fashion trends that peaked in the 1970s. Batik is vulnerable to fashion cycles and needs to adapt to changing circumstances. Tourism may be one of the means of keeping these craft skills alive, particularly if linked to what is increasingly being called cultural tourism. It remains uncertain, however, whether or not tourists are prepared to pay well for quality products that are not available in the countries whence they came. As Tri Budhi Sastino notes, crafts serve both the tourist and export markets and it remains unclear whether or not the availability of similar crafts in the tourists' home countries will deter spending during the vacation. Handicrafts of the kind found of roadside stalls in Bali are readily available in shops and markets in Europe and North America.

Several contributors (Tjoa and Wagner, Sulistyawati, van Oss) stress the importance of the Internet in stimulating demand and raising consciousness. The interpretation of crafts that are not likely to be well understood by visitors should also include well trained guides and informative guidebooks. Outside the specialist tours, however, one should not overestimate the importance of handicrafts. From a survey referred to by Sulistyawati it would appear that textiles alone do not necessarily act as a draw, though they may contribute to the generalised image of a destination.

The tension between cloth as a commodity and cloth as a sacred artefact is a recurring theme. Batik , according to Oetari Siswomihardjo, continues to worn as part of official dress, but is also ritually significant. Sekimoto reminds us that batik has long been a cultural artefact as opposed to a commodity and is in Japan closely associated with Indonesian identity. Debates about tradition versus modernity, however, have been around a long time and what many authors suggest is that it is not particularly helpful in the case of batik to argue that one necessarily negates the other. Nevertheless, traditional textiles are continuing to be turned into commodities and new designs are being created to satisfy new markets. Not only are the designs affected by these changes, but so are the social relations of production. As Smedjebacka notes, changes in occupational roles associated with gender may also accompany the process of commoditization.

Tourism as one of the forces of globalization undoubtedly contributes to this process of cultural adaptation and many authors grapple with the apparent contradiction of the need to earn income on one hand and the desire to protect the dignity of traditional cultures on the other. Craft tours can also take tourism to areas that want the revenue from visitors but are not directly linked to international tourism. Poor areas also have to face stark choices concerning employment, but Crippen notes the value of handicrafts in raising self esteem among rural women in India. In Chiang Mai in Thailand a weaving career may provide an alternative to sex tourism. Several contributors try to resolve some of these contradictions by separating the sacred (traditional cloth) from the profane (tourism), often by physically keeping tourism apart from the more important ritual activities. The application of Goffman's celebrated distinction between front-stage and backstage (Allcock, 1995, p. 110) is controversial in tourism, but clearly indicates a potential area of further research. Graburn offers some useful observations with regard to his research with the Inuit. The indigenous people of the Canadian and American Arctic were drawn into the world system as trappers and only later did tourism become important. They produce souvenirs to satisfy the following quite different demands. First, they provide buyers with attractive arts and crafts. Second, they strive to control the outside world's image of themselves. Third, they aim to be commercially viable in an increasingly competitive world.

References

Allcock, J.B. (1995), 'International tourism and the appropriation of history in the Balkans' in M.-F. Lanfont (ed.), *International Tourism: Identity and Change*, London, Sage.

Hart, K. (1982), 'On commoditization' in E.N. Goody (ed.), *From Craft to Industry: The Ethnography of Proto-Industrial Cloth Production*, Cambridge, Cambridge University Press.

Printed and bound by CPI Group (UK) Ltd, Croydon, CR0 4YY

21/10/2024

01777083-0002